# Biotechnology;
# Building on Farmers'
# Knowledge

*Edited by:*

Joske Bunders, Bertus Haverkort and
Wim Hiemstra

MACMILLAN

First published 1996 by
MACMILLAN EDUCATION LTD
London and Basingstoke
*Companies and representatives throughout the world*

ISBN 0–333–67082–5

| 10 | 9 | 8 | 7 | 6 | 5 | 4 | 3 | 2 |
|----|----|----|----|----|----|----|----|----|
| 06 | 05 | 04 | 03 | 02 | 01 | 00 | 99 | 98 |

Typeset by EXPO Holdings, Malaysia

This book is printed on paper suitable for recycling and made from fully managed and sustained forest sources.

Printed in Hong Kong

A catalogue record for this book is available from the British Library.

Cover illustration by Kees Manintveld. See also p86.

*Related title*

FARMING FOR THE FUTURE Published by Macmillan and ILEIA: by Coen Reijntjes, Bertus Haverkort and Ann Waters-Bayer examines the strategies and techniques of Low-External-Input and Sustainable Agriculture (LEISA) in the tropics. It is based on eight years' work by the Information Centre for Low-External-Input and Sustainable Agriculture (ILEIA) in conjunction with the ETC Foundation in the Netherlands. With the aid of its network of over 10 000 members, ILEIA has collected the experiences of innovative farmers, fieldworkers and supporting scientists in developing productive and sustainable forms of agriculture. The scientific principles behind the various LEISA systems and techniques have been analysed, with the advisory support of staff members from Wageningen Agricultural University and independent professionals.

The central concern of the book is how development workers can assist small-scale farmers in making the best use of low-cost local resources to solve their agricultural problems. Emphasis is therefore on methods of Participatory Technology Development (PTD) to find site-specific solutions and to raise the overall productivity of farming in a sustainable way.

# Contents

# About the authors

**Joske F.G. Bunders** graduated in chemistry and physics. She wrote her Ph.D. thesis on participatory approaches to the development of science-based innovations in agriculture at the Vrije Universiteit Amsterdam. She is currently Head of the university's Department of Biology and Society. She has supervised numerous projects involving biotechnology for small-scale farmers, notably in Zimbabwe, Ghana, Pakistan, Bolivia and Bangladesh. *Address:* Faculty of Biology, Vrije Universiteit Amsterdam, de Boelelaan 1087, 1081 HV Amsterdam, Netherlands.

**Bertus Haverkort** has considerable working experience in sustainable agriculture and participatory technology development in developing countries, having worked in Colombia, Thailand, Ghana and several other countries in Africa and Latin America. He has served as programme manager at the Information Centre for Low-external-input and Sustainable Agriculture (ILEIA) and is now manager of the ETC Foundation. He is particularly interested in farmers' indigenous knowledge and coordinates COMPAS, a project to compare and support indigenous agricultural systems. *Address:* ETC Foundation, Kastanjelaan 5, P.O. Box 64, 3830 AB Leusden, Netherlands.

**Wim Hiemstra** has worked with ILEIA since 1984, editing the *ILEIA Newsletter* and other publications. He has field experience in Indonesia, Ghana and Thailand, and has visited and supported many non-government organisations (NGOs), mainly in Asia. He is part of ILEIA's Action Research team and supports the Third World Working Group of the International Federation of Organic Agriculture Movements (IFOAM), and is also working on COMPAS. He is particularly interested in the North-South dialogue on sustainable agriculture. *Address:* ETC Foundation, Kastanjelaan 5, P.O. Box 64, 3830 AB Leusden, Netherlands.

**Saleem Ahmed** began his career at the Department of Geology, University of Karachi, before joining Exxon (Pakistan), a private-sector input supply company, as technical services advisor. He is currently a Senior Fellow at the East-West Center, Hawaii, and a consultant for the Asian Development Bank, the United Nations Development Program, the Government of Pakistan and several private-sector companies. His work experiences cover a wide range of policy issues in sustainable agriculture, including the development of biopesticides. He has also trained graduate students and young professionals. He has twice been awarded a Fullbright fellowship, for studies on organic farming in Japan and botanical pest control in India. *Address:* East-West Center, 1777 East-West Road, Honolulu, Hawaii 69848, USA.

**Helen Appleton** is a policy researcher in the Intermediate Technology Development Group. *Address:* Intermediate Technology Development Group, Myson House Railway Terrace, Rugby, CV21 3HT, England.

**Mike Battcock** is the International Programme Manager for Agro-processing in the Intermediate Technology Group. He has over 10 years experience in the development of successful food processing activities in Asia, Africa and Latin America, particularly on small scale enterprise training in Bangladesh, spice processing in Sri Lanka and drying fruits in Pakistan. *Address:* Intermediate Technology Development Group, Myson House Railway Terrace, Rugby, CV21 3HT, England.

**Trygve Berg** graduated in Agronomy at the Agricultural University of Norway in 1970, and did a Ph.D. in plant breeding at the same place in 1975. He worked with agronomy research in Norway until 1982 and in the Sudan from 1982 to 1986. Since 1992 he has been employed at the Centre for International Environment and Development Studies at the Agricultural University of Norway. His main interest is the management of plant genetic resources and participatory breeding methods. *Address:* Centre for International Environment and Development Studies, Noragric, P.O. Box 5001, N-1432 Ås, Norway.

**Walter S. de Boef** was trained at Wageningen Agricultural University as a breeder and rural sociologist. He works for the CPRO-DLO Centre for Genetic Resources, the Netherlands. He coordinated the initial phase of the Community Biodiversity Development and Conservation (CBDC) Programme, a development and research programme of NGOs, National Agricultural Research Stations (NARS) and gene banks in eighteen countries. Presently, he is involved in the development of approaches of in-situ conservation of agro-biodiversity, participatory plant breeding and local seed supply systems. *Address:* Centre for Genetic Resources, P.O. Box 16, 6700 AA Wageningen, Netherlands.

**Jacqueline E.W. Broerse** graduated in medical biology at the Vrije Universiteit Amsterdam in 1988, after which she worked at the Dutch Directorate General for International Cooperation for two years on the establishment of a Dutch policy and programme in the field of biotechnology and development. Currently, she is a lecturer at the Department of Biology and Society of the Vrije Universiteit Amsterdam. Her research focuses on strategies for generating and diffusing biotechnological innovations in the agricultural sectors of developing countries. *Address*: Faculty of Biology, Vrije Universiteit Amsterdam, de Boelelaan 1087, Amsterdam 1081 HV, Netherlands.

**Vishaka Hidellage** is the Country Programme Manager of the Agro-Processing Programme in Intermediate Technology Sri Lanka. The programme has a successful small enterprise training programme and is working on semi-processing of tomatoes and small scale processing of cashews. *Address:* Intermediate Technology Sri Lanka, 15-B Alfred's Place, Colombo 3, Sri Lanka.

**Andrew Jones** is a food scientist who, between 1985 and 1995 worked for the Intermediate Technology Development Group on small scale food processing technologies and small enterprises development. He has worked for nearly four years in Sri Lanka and for one year in Peru. He now works for The Body Shop International in the Fair Trade department with a responsibility to find more suppliers of raw materials according to the company's Fair Trade criteria. *Address:* The Body Shop International PLC, Watersmead, Littlehampton, Sussex BN17 6LS, England.

**Anne Loeber** is a political scientist, and has trained in social anthropology and public administration. She worked on new approaches to directing technology development, focusing on methods for facilitating exchange of information between users and producers, especially in the field of biotechnology. Her current research, at the Department of Political Science, University of Amsterdam, focusses on the role of interactive processes of innovation and technology assessment in influencing technological and institutional change towards more sustainable development. *Address:* Department of Political Science, University of Amsterdam, O.Z. Achterburgwal 237, 1012 DL Amsterdam, Netherlands.

**Stéphane Malo** received a B.Sc. in sociology at the Université de Montréal and studied economics at Concordia University. Following this, he completed an MA in sociology at the Université de Quebec à Montréal. In 1992, he joined the Maastricht Economic Research Institute on Innovation and Technology (MERIT) for the MA programme in Innovation Management and Technology Policy. After a brief stay at the department of Biology and Society of Vrije Universiteit Amsterdam, he returned to MERIT as a Ph.D. candidate. He actively participates in MERIT research projects on Intellectual Property Rights issues, regulatory policies and strategic partnerships in the field of biotechnology. *Address:* MERIT, P.O. Box 616, 6200 MD Maastricht, Netherlands.

**Evelyn Mathias** holds a Dr. Med. Vet. from the University of Giessen, in Germany and an MS in international development from Iowa State University (ISU). She has conducted field research on goat feeding behaviour in Tunisia, on the reproductive physiology of swamp buffalo in Thailand and on ethnoveterinary medicine in Indonesia. She has also served as visiting lecturer at Bogor Agricultural University in Indonesia, research associate with the Center for Indigenous Knowledge for Agriculture and Rural Development (CIKARD) at ISU, and as coordinator of the Regional Program for the Promotion of Indigenous Knowledge in Asia at the International Institute of Rural Reconstruction (IIRR), in the Philippines. *Address:* Grubenfeld 7, 51467 Bergisch Gladbach, Germany.

**Constance M. McCorkle** has a Ph.D. in anthropology from Standford University, California, and currently works as an independent researcher, author and public speaker in her capacity as president of her own consulting firm. Recently, she has served as Director of Research and Evaluation for the Gender in Economic and Social Systems (GENESYS) Project of the United States Agency for International Development (USAID) and as Director of USAID's in-house staff environmental training programme, both in Washington D.C. Before that she was a faculty member at the University of Missouri-Columbia's Department of Rural Sociology, where she coordinated the Sociology Project of the Small Ruminant Collaborative Research Support Program. She has worked in more than 20 developing countries, with long-term studies in Ghana, Mali, Peru and Burkina Faso. *Address*: CMC Consulting, 7767 Trevino Lane, Falls Church, Virginia 22043, USA.

**Michael N. Kibue** is a graduate of Nairobi University, where he read environmental sciences. He owns a two-acre farm in Kiambu District of Central Province. He has divided his farm into two, practising traditional farming in one half and modern farming in the other. His aim is to determine which system is more profitable. *Address*: P.O. Box 62802, Nairobi, Kenya.

**Guido Ruivenkamp** is senior lecturer for the Working Group Technology and Agrarian Development at Wageningen Agricultural University since May 1993. From 1984 until 1993 he did Ph.D. research on "The introduction of biotechnology into the agro-industrial chain of production; changing over to a new form of labour organisation" at the University of Amsterdam. From 1989 until 1990 he was senior research officer at the Centre for Technology and Policy Studies (STB) of the Netherlands' Organisations for Applied Scientific Research (TNO). From 1990 until 1993, he was director of the Centre of Agriculture and Biotechnology, research department of the Western Farming and Horticulture Organisation. *Address:* Working Group Technology and Agrarian Development, Department of Irrigation and Soil and

Water Conservation, Nieuwe Kanaal 11, 6709 PA Wageningen, Netherlands.

**Gaby Stoll** did an M.Sc. and a Ph.D. in agricultural biology at the University of Hohenheim in Stuttgart, Germany before working for two years with the Networking and Information Centre for Sustainable Agriculture in the Third World (AGRECOL), based in Switzerland. She has spent three years in Thailand, coordinating farmers' research on plant-based biopesticides, and two years in the Philippines, where she was a research associate at the Philippine-German Biological Plant Protection Project. Since 1994, she has been programme officer at Misereor, a German co-financing organisation. *Address*: Maria-Theresia Allee 265, 52074 Aachen, Germany.

**Theo van de Sande** graduated in political science (international relations) at the University of Nijmegen. He has been a research fellow at the University of Twente and the Vrije Universiteit Amsterdam, where he has studied the socio-economic aspects of technology transfer and technology policy in developing countries, with special emphasis on biotechnology. He is currently working at the Dutch Directorate General for International Cooperation for its Special Programme Biotechnology and Development Cooperation. *Address*: Ministry of Foreign Affairs, Directorate General International Cooperation, P.O. Box 20061, 2500 EB The Hague, Netherlands.

**Bert Visser** studied Molecular Sciences at the Agricultural University of Wageningen and obtained his Ph.D. at the Faculty of Medicine, University of Utrecht. He then joined the Centre for Plant Breeding and Reproduction of the Research Department DLO of the Netherlands' Ministry of Agriculture, Nature Management and Fisheries. As a senior scientist in the molecular biology department, he started research on insect resistance in transgenic plants. In 1992 he joined the Special Programme Biotechnology and Development of the Ministry of Foreign Affairs, responsible for institutional development. He recently returned to DLO and now works with the Institute of Animal Science and Health, where he is responsible for the assessment of licence applications for veterinary medicines and feed additives, marketed in the Netherlands. *Address:* Institute of Animal Science and Health ID-DLO, P.O. Box 65, 8200 AB Lelystad, Netherlands.

**Helen Wedgewood** is a social scientist for the Food Security Programme of the Intermediate Technology Development Group. She has worked in integrated rural development in Africa, food systems research in South East Asia for the FAO and as a lecturer in agricultural economics at Wye College in England. *Address:* Intermediate Technology Development Group, Myson House Railway Terrace, Rugby, CV21 3HT, England.

# Preface

*Prof. M.S. Swaminathan*

In his message to the International Commission on Peace and Food, the Secretary General of the United Nations, Mr Boutros Boutros Ghali wrote, "As development becomes imperative, we are faced with the necessity of giving new meaning to the word. Reflecting on development is thus the most important intellectual challenge in the coming years." (Report of ICPF, Zed Books, London, 1994). If we draw a balance sheet of the gains and problems associated with the developmental pathways adopted since the end of World War II in 1945, we can see a number of achievements, but also a number of concerns.

On the one hand we see an impressive progress in science and technology, such as information technology leading to unprecedented economic growth and unusual opportunities for food, health, literacy and jobs. The end of the cold war provides the opportunity to spend less on defence and more on development and helps in the spread of democracies and renewed faith in UN and other multilateral institutions. Skin-colour based apartheid has come to an end.

Major concerns relate to the indicators measuring environmental and social sustainability; damage to life support systems and potential adverse changes in climate and sea level; economic and gender inequity; unemployment and social disruption due to disparities in entitlements; growing violence and ethnic strife; AIDS; debt burden and emerging technological and economic apartheid.

Among the problems which affect contemporary human development the most significant are: poverty, unemployment, hunger, ignorance, disease, squalor and exclusion. The World Commission on Culture and Development in its report titled, "Our Creative Diversity" (UNESCO, 1995) has pointed out that if we balance information and knowledge with wisdom, rights with duties, and ends with means, a new Renaissance; a new creative vision of a better world, awaits us. Such a Renaissance will need for its initiation the widest possible democratic mobilisation of ideas, knowledge and skills. It is in such a context, that this book "Biotechnology; Building on Farmers' Knowledge" is a timely and welcome one.

The editors, Joske Bunders, Bertus Haverkort and Wim Hiemstra have defined biotechnology as: "the application of indigenous and/or scientific knowledge to the management of (parts of) micro-organisms, or of cells and tissues of higher organisms, so that these supply goods and services of use to human beings". Through this definition, they have brought out the underlying philosophy of this book, namely that knowledge is a continuum and that success lies in strengthening and not severing the links in the innovation chain evolved over the millennia, thus learning lessons from both low external input agriculture and high external agriculture. Finding the balance is the challenge.

The book appropriately starts with a farmer's view and then proceeds to cover different aspects of animal health, biopesticides, food processing and crop genetic resources. It then proceeds with an assessment of science-based biotechnology and concludes with a chapter on building on farmers' practices.

I am glad a chapter on the socio-political context of recent developments in the area of biodiversity and biotechnology has also been included. From the early origins of agriculture about 12,000 years ago and until now, agricultural research, with particular reference to plant and animal introductions and breeding, has progressed in an environment of a free flow of ideas and information and exchange of material. This situation has however changed from 1 January 1995, when the World Trade Agreement and the associated Trade-Related Intellectual Property Rights (TRIPS) regime came into force. Mutual suspicion is growing between countries rich in biodiversity and those which are rich in modern biotechnological expertise. Cooperation is getting replaced with conflicts and more investment is being made on lawyers than on breeders. The casualty will be our hope for a food secure world.

The concluding chapter on developing appropriate biotechnologies for poor rural people is therefore a fitting finale to this timely book. The authors show in this chapter how we can include those who have so far been excluded from deriving economic benefits from their knowledge and genetic conservation ethics. How can we recognise and reward the collective wisdom based on real-life experience of communities, in contrast to the innovations of individuals? The legal instruments essential for recognising the intellectual property contributions of communities are yet to evolve. This book shows the way to keep biotechnology research and development in the realm of a "public good" activity. For this purpose, the authors recommend a participatory and interactive methodology of research involving farmers, scientists and policy makers, as well as a broader process of institutional change. This is the way to proceed, if we are to follow Albert Einstein's advice. "Concern for man himself and his fate must always form the chief interest of all techniques and endeavours in order that the creation of our minds shall be a blessing and not a curse". The world today needs an evergreen revolution, where we can achieve continuous improvements in biological productivity without associated ecological harm. Also, the concept of sustainability should not be restricted only to ecological and economic concerns but should also include the social dimension.

As we approach a new century, we are confronted with the task of producing more and more food and other agricultural commodities from less and less per capita arable land and irrigation water resources. Sustainable food security can be achieved only through higher productivity per units of land, water, energy and time. All this calls for the integration of the best in traditional ecological wisdom and prudence and modern technologies.

The "green revolution" technologies of the sixties and seventies were scale neutral with reference to their relevance to farmers operating different sizes of holdings. They were however not resource neutral, since more market purchased inputs were needed for higher output. Biotechnology offers opportunities for integrating resource neutrality with scale neutrality in technology development and dissemination. This is why a movement for building on indigenous biotechnology is an urgent necessity if the concept of sustainable agriculture is to become a reality at the field level.

M.S. Swaminathan
Madras, India
30 January 1996

# Foreword

This book is the result of the many activities initiated and sponsored by the Special Programme on Biotechnology and Development Cooperation for the Netherlands' government. The five-year Programme was established in 1992 by the Netherlands' Minister for Development Cooperation. It aims at improving the access of developing countries to biotechnological expertise and innovation with a focus on using biotechnology for the benefit of small scale farmers and producers. To achieve this we concentrate on: technical collaboration with four developing countries (Kenya, Zimbabwe, Colombia and India); initiation and support of relevant international initiatives and networks, including biosafety, intellectual property and information systems; and the integration of development in biotechnology policies of the Netherlands' government. What is fundamental to the country programmes, and unique, is the participatory bottom-up process applied, starting with local needs' assessments and priority setting through to project formulation and implementation. In this process end-users, researchers, policy makers and NGOs participate. High priority is also given to the transfer of ownership to the countries.

During the initial phase of the Biotechnology Programme the Minister's Biotechnology Advisory Committee discussed possible subject areas for improvement of modern biotechnology especially the area of indigenous (bio)technology. Therefore ETC-ILEIA was invited to organise a competition. The Biotechnology Advisory Committee selected ten winners from more than 40 contributions sent in by ILEIA Newsletter readers.

It was then decided to produce a book from the wealth of information received. The result is a unique compilation of rural biotechnologies and their potential. It complements the identified priorities in the collaborative country programmes of the Biotechnology Programme and is therefore highly recommended.

I would like to extend my gratitude to all collaborators to this publication, especially ETC and the Vrige Universiteit Amsterdam.

Th. J. Wessels
Head of the Biotechnology Programme
DGIS, Ministry of Foreign Affairs
Netherlands

# Acknowledgements

This book is a synthesis of the experiences of many people: farmers, field workers and researchers, all of whom were prepared to document their knowledge and to share it with others. We are profoundly grateful to them.

Among the end users of biotechnology are many farmers in developing countries who live and work under difficult conditions in areas that are often marginal for agriculture, remote from markets and lacking in basic infrastructure and services. A major challenge for Third World agricultural research and development is to generate and disseminate products and services of value to such people. To this end in the context of the Netherlands' Special Programme on Biotechnology an inventory was made of farmers' existing practices in biotechnology through an international contest.

The contest unearthed a wealth of relevant material. Especially in the domain of food processing, rural women have a vast stock of knowledge, little of which is documented. First prize went to Hamid A. Dirar, of the Faculty of Agriculture at the University of Khartoum, for his paper entitled "The Indigenous Fermented Foods and Beverages of Sudan". Martin Anike, of NTE Farms in Nigeria; Davidson and Alice Mwangi, of Amaranth and Natural Foods in Kenya; and David Syakacha, in Zambia, won second prizes. Third prizes were awarded to C.I. Ezedinma and I. Igbinnosa of the International Institute for Tropical Agriculture (IITA) in Nigeria; E. Manzungu of Chirdzi Research Station in Zimbabwe; F.K. Sanyu of the Agricultural Extension Service in Uganda; C. Berhe of the Environment and Development Society of Ethiopia; A.O. Ambimbola of the University of Ilorin in Nigeria; and M.N. Kibue, a farmer from Kenya.

Later it was decided to compile these contributions into a book and to elaborate on the topic. We identified a dozen authors prepared to write on different aspects of biotechnology. This was not an easy task, as reliable information often was not available. Where it was, it had to be processed and synthesized into a format that would lend the book consistency. Many contributions had to be rewritten several times before their authors and the editors were satisfied with them. To all our contributors we express our gratitude for their patience and hard work, and we hope they like the final result.

We greatly appreciated the work of the external reviewers who commented on early drafts of several chapters. Chapter 5, on food processing, was read by Robert Nout, Lecturer in Food Microbiology, Department of Food Science, Wageningen Agricultural University, Netherlands. Daniela Soleri, of the Centre for People, Food and Environment in Tucson, Arizona, USA took on the difficult Chapter 6, on the controversial subject of crop genetic resources. This chapter was also reviewed by Jaap Hardon, plant breeder and Director, Centre for Genetic Resources, Wageningen, Netherlands. No less than four reviewers were invited to comment on Chapter 7, which discusses the potential of science-based biotechnology. They were: Willem Siekema, biotechnologist with the Centre for Plant Breeding and Reproduction Research at the Agricultrual Research Department of the Dutch Ministry of Agriculture, Nature and Fisheries; Robert Paling, doctor of veterinary medicine at the Office for International Cooperation, Faculty of Veterinary Medicine at the University of Utrecht, Netherlands; Robert Nout and Jaap Hardon. Chapter 8, on the social and political context of biotechnology, was reviewed by Peter Commandeur, Robin Pistorius and Jeroen van Wijk, editors of *Biotechnology and Development Monitor,* Amsterdam, Netherlands. Chapter 9, on the technology development process, was read by Laurens van Veldhuizen, consultant in extension education with the ETC Foundation, and by Ann Waters-Bayer, consultant in rural sociology with ETC.

Despite what some would call a surfeit of reviewing, errors of fact or interpretation may remain on what are often controversial, or at least extremely complex, issues. For these and any other blemishes, the editors and authors accept full responsibility.

We would also like to express our sincere appreciation to the Special Programme on Biotechnology and Development Cooperation of the Directorate General for International Cooperation at the Dutch Ministry of Foreign Affairs, for its interest in this subject, its confidence in the ETC Foundation and Vrige Universiteit and for its contribution in the funding of the book.

Lastly, we would like to thank Simon Chater, who not only helped in correcting the language and in structuring the book but also pointed out inconsistencies in early drafts. In doing so he helped bring balance in the book.

Joske Bunders, Bertus Haverkort and Wim Hiemstra
Leusden,
29 March 1996

# Abbreviations

| | |
|---|---|
| AFLP | amplified fragment length polymorphism |
| AGRECOL | Networking and Information Centre for Sustainable Agriculture in the third world |
| APONET | International Network on Apomixis Research |
| BGA | blue-green algae |
| BNF | biological nitrogen fixation |
| BPC | botanical pest control |
| BST | bovine somatotropin |
| Bt | Bacillus thuringiensis |
| CBDC | Community Biodiversity Development and Conservation Programme |
| CBPP | contagious bovine |
| CCPP | caprine pleuropneumonia |
| CGIAR | Consultative Group on International Agricultural Research |
| CIAT | Centro Internacional de Agricultura Tropical |
| CIKARD | Center for Indigenous Knowledge for Agriculture and Rural Development |
| CIMMYT | Centro Internacional de Mejoramiento de Maïz y Trigo |
| CIP | Centro Internacional de la Papa |
| ELISA | enzyme-linked immuno-sorbent assay |
| EMBRAPA | Empresa Brasileira in de Pesquisa Agropecuaria |
| EPA | Environmental Protection Agency |
| ET | embryo transfer |
| ETC | Educational Training Consultants |
| EVM | ethnoveterinary medicine |
| FAO | Food and Agricultural Organisation |
| FMD | foot-and-mouth disease |
| FSH | follicle-stimulating hormone |
| GATT | General Agreement on Tariffs and Trade |
| GENESYS | Gender in Economic and Social Systems Project |
| GSP | Generalized System of Preferences |
| HEIA | high-external-input agriculture |
| HFCS | high fructose corn syrup |
| IARCs | international agricultural research centres |
| IBU | interactive bottom-up |
| ICARDA | International Centre for Agricultural Research in the Dry Areas |
| ICPF | International Commission on Peace and Food |
| ICRISAT | International Crops Research Institute for the Semi-Arid Tropics |
| IDS | Institute of Development Studies (Sussex) |

| | |
|---|---|
| IFOAM | International Federation of Organic Agriculture Movements |
| IIMI | International Irrigation Management Institute |
| IIRR | International Institute of Rural Reconstruction |
| IITA | International Institute of Tropical Agriculture |
| ILCA | International Livestock Centre for Africa |
| ILEIA | Information Centre for Low-external-input & Sustainable Agriculture |
| ILRAD | International Laboratory for Research on Animal Diseases |
| IPM | integrated pest management |
| IRRI | International Rice Research Institute |
| ISNAR | International Service for National Agricultural Research |
| ITDG | Intermediate Technology Development Group |
| KRIBP | Krishak Bharati Cooperative Indo British Rainfed Farming Project |
| LARC | Lumle Agricultural Research Centre |
| LEIA | low-external-input agriculture |
| MAb | monoclonal antibody |
| MERIT | Maastricht Economic Research Institute on Innovation and Technology |
| MIRCENs | Microbiological Resources Centres |
| NARS | National Agricultural Research Stations |
| NGO | non-government organisation |
| NifTAL | Nitrogen Fixation for Tropical Agricultural Legumes |
| OPVs | open-pollinated varieties |
| PAR | participatory action research |
| PBRs | plant breeders' rights |
| PCR | polymerase chain reaction |
| PGPR | plant-growth-promoting rhizobia |
| PGR | plant genetic resources |
| PPA | Plant Patent Act |
| PRA | participatory rapid appraisals |
| PSM | phosphate-solubising micro-organisms |
| PST | porcine somatotropin |
| PTD | participatory technology development |
| PVPA | Plant Variety Protection Act |
| RFLP | restriction fragment length polymorphism |
| RAPD | random amplified polymorphic DNA |
| SMIP | Sorghum and Millet Improvement Program |
| SSF | solid state fermentation |
| STB | Centre for Technology and Policy Studies |
| STOA | Scientific and Technological Options Assessment |
| TREE | Technologies for Rural Ecological Enrichment |
| TRIPS | Trade-Related Intellectual Property Rights |
| UPASI | United Planters' Association of South India |
| UPOV | International Convention for the Protection of New Plant Varieties |
| USAID | United States Agency for International Development |
| VA | vesicular-arbuscular |

# 1 Introduction

*Joske Bunders, Bertus Haverkort and Wim Hiemstra*

## What is biotechnology?

During the 1970s the word biotechnology, which had previously had a career in food processing and agro-industry, began to be used by the Western scientific establishment to describe a relatively narrow range of laboratory-based techniques then being developed in biological research. In fact the term could, perhaps should, be understood in a much broader sense to connote the whole range of methods, both ancient and modern, used to manipulate organic matter to meet human needs. For the purposes of this book we therefore define biotechnology as: The application of indigenous and/or scientific knowledge to the management of (parts of) micro-organisms, or of cells and tissues of higher organisms, so that these supply goods and services of use to human beings.

Thus understood, biotechnology is a continuum of technologies ranging from the simple to the sophisticated and from those long established and widely applied by ordinary people to those more recently developed and, as yet, comparatively little used except by highly trained specialists. Agricultural applications cover fields as diverse as traditional fermentation technology in food processing and the use of a particle gun to transfer genes from one plant species to another. A mother in her kitchen trying out a new way of making bread is as much a biotechnologist as her daughter who goes out to work at the laboratory on her Ph.D. in the genetic engineering of a new tomato plant.

## The global context

Biotechnology research and development is conducted in a world more than ever in need of its products. In 1995, over 700 million people did not have sufficient food for a healthy and productive life (Pinstrup-Andersen and Pandya-Lorch, 1994). Up to a billion survive on less than a dollar a day (Brown et al., 1994).

When agriculture began, around the tenth century BC, the doubling time for the world's human population was 8000 years. Now it is a mere 40 years. By 2025, our planet will contain about 8.5 billion people, over three-quarters of whom will live in the

1

developing countries. Global food production will have to rise by 2.6 billion tonnes just to maintain current per capita food consumption. If diets are to improve among the world's poor and hungry, an extra 4.5 billion tonnes of food will be needed.

The outlook for Asia and sub-Saharan Africa is particularly grim. Unless current trends alter, food production in 2025 in these two regions will fall 40 per cent and 60 per cent short of needs respectively. But the world's other developing regions; West Asia, North Africa, Latin America and the Caribbean, and the Pacific and Indian Ocean islands, also give cause for concern.

Although food production increased by 39 per cent in the South during the 1980s, in most countries the increase barely kept pace with population growth. In Africa, per capita food production fell during the decade. In this region, most gains in production came from increases in the area cultivated rather than higher yields. In Asia, yields rose steadily in the 1970s and the first half of the 1980s, but have since stood still or fallen slightly.

Further expansion of the area cultivated is impossible in most of the world's developing regions. This means that future gains in food production will have to come from increased yields. Technological change, generated through research, will be vital in bringing about such increases.

## Two agricultures

In analysing the potential of biotechnology, we distinguish between two broad types of agriculture: low-external-input agriculture (LEIA) and high-external-input agriculture (HEIA).

The practitioners of HEIA concentrate on the specialised production of a few commodities, using technologies that enable genetically homogeneous seeds and animals to be kept under conditions that allow maximum output. To control the production environment, microclimates and water regimes are managed and agrochemicals such as fertilisers and pesticides are applied. This type of agriculture, which is strongly market oriented, tends to be capital-intensive and highly mechanised, consuming large quantities of non-renewable resources such as oil and phosphates. It is only possible where ecological conditions are relatively uniform and can easily be controlled (for example in irrigated areas) and where delivery, extension, marketing and transport systems are good.

Throughout the 1970s and during the first half of the 1980s, HEIA led to rapid gains in food production in the high-potential (mostly irrigated) areas of Asia and Latin America. In recent years, however, yields have stagnated or declined in many areas, while the environmental costs of HEIA have become steadily more apparent. Besides the profligate consumption of non-renewable resources, these costs include various disturbances to agroecosystems, especially the destruction of natural enemies of crop pests, the pollution of soil and water resources, the entry of toxins into the food chain and negative effects on human health.

LEIA, which is the main focus of this book, is the predominant form of agriculture in the large proportion of the developing world in which the physical environment, the commercial infrastructure and price ratios do not allow the widespread use of purchased inputs (Reijntjes et al., 1992). Wolf (1986) estimates that some 1.4 billion people currently depend on LEIA for their livelihood. LEIA, which is the prevalent form of agriculture throughout sub-Saharan Africa, is found in the rainfed, undulating lowlands of the humid and subhumid tropics, in the drylands, in the highlands and in forested areas. Many areas in which LEIA is practised have fragile or problematic soils and unpredictable water supplies.

The area under LEIA is growing. Under structural adjustment programmes, external inputs have become more expensive, with the result that fewer farmers can afford them. As levels of pest resistance rise, even relatively affluent farmers in high-potential areas are cutting back on their use of pesticides.

In order to survive, many farmers in LEIA systems are forced to exploit the land beyond its carrying capacity. This leads to accelerated deforestation, soil degradation and vulnerability to pest attacks, torrential rain or extended drought. Many tropical land use systems are undergoing a downward spiral of loss of vegetative cover, nutrient depletion and soil erosion, accompanied by economic, social and cultural disintegration. The sustainability of agriculture in these marginal lands poses a special challenge to those responsible for it.

## Two research paradigms

Much of the technology generated through formal research has benefited HEIA. Researchers themselves acknowledge that their efforts have been less successful in LEIA systems, where relatively few innovations have been adopted by farmers.

Debate on the reasons why resource-poor farmers do not adopt new technology began in the early 1970s, when the first socio-economic studies on the impact of the Green Revolution noted that early adopters tended to be larger, wealthier farmers. Later it became apparent that, in irrigated areas, smaller farmers, though more cautious, also adopted, in order to stay competitive. But the new technology, heavily reliant on the use of external inputs, had failed to take off altogether in the rainfed areas.

At first it was suspected that the problem might be information: farmers in rainfed areas did not know about promising technologies, or at least did not know enough about them. Extension services in such areas were less well developed than in irrigated areas, with the result that the message was not getting through. Then it became clear that ignorance was not the problem.

Farmers were in fact extremely rational in their decision making, but they applied different criteria to those of researchers for assessing a technology's usefulness. The technologies effective in HEIA systems were simply not appropriate for the more diverse and complex systems found in LEIA areas.

As a result, people and institutions began experimenting with new approaches to developing technologies. Broadly, two movements can be distinguished, one within the formal research system, and one operating largely outside it. In both cases, many valuable lessons have been learnt.

In the formal system, scientists sought to improve their understanding of farmers' constraints, both at the farm (micro) level and in the wider policy and institutional environment (macro) level. Social scientists were brought into the research process to analyse these constraints. Beginning in the early 1970s, farming systems' research emerged as a new multidisciplinary approach in which scientists tried to take whole production systems into account rather than focussing on selected aspects of them. For the first time, farmers were included in the research process, although more often as passive providers of land, labour and information than as active contributors to the design of new technology. Diagnostic surveys, often mounted on a large scale, became the tool through which information was gathered. Social sciences, such as crop sociology and anthropology, were asked to generate the missing information. The focus of research broadened to include first livestock then trees and other natural resources alongside conventional crop production.

Early experiences revealed the difficulties of conducting multidisciplinary research in agriculture. Biological and social scientists were often not familiar with each others' work and did not know what to ask of each other. This was exacerbated by differences in language (jargon) and thinking. Another common problem was that the input of social scientists was often left to the end of the research cycle, when they were asked to help fine-tune technology, facilitate its transfer and assess its impact. By that stage it was usually too late to influence technology design. These and similar problems are explored in McCorkle (1989).

During the 1980s, formal researchers came under growing pressure to increase the cost-effectiveness of research. The diagnostic phase of the systems approach began to seem too protracted, so various short-cuts, such as rapid rural appraisal, were developed. Over the same period the development of new production systems, which farmers were initially expected to adopt in their entirety, gave way gradually to a less rigid approach in which farmers were presented with a range of technology options from which they could select. Social scientists began to be included at all phases of the research process, especially the diagnostic phase. There was growing emphasis on environmental and equity issues, increasingly recognised as equal in importance to the challenge of increasing production.

Meanwhile, a more fundamental constraint had come to the fore: even the information collected by social scientists might not provide sufficient insight into farmers' constraints. The way in which sociological information is sought and assessed can easily lead to misinterpretation by the researcher. If there is no corrective mechanism, the value of the information is likely to be low (Chambers and Ghildyal, 1985; Uphoff, 1992). The solution to this problem lay in the more active participation of the subjects of research, resource-poor farmers themselves. If their knowledge and skills were to be tapped by formal research and development, farmers had to be involved in decision-making throughout the project. Thus farmer participation became the new buzz phrase in the formal research system.

In 1988, the International Service for National Agricultural Research (ISNAR) launched a project to analyse the experiences of nine national institutes in developing countries with what it called on-farm client-oriented research. The conclusions of this study were sobering, most formal research institutes had experienced severe problems in applying client-oriented approaches. Among them were difficulties in maintaining a multidisciplinary focus, subversion of the research process by internal and external attempts to exert influence, a loss of early enthusiasm and a tendency to fall back on old, proven, methodologies. In many cases, the term farmer participation had been used when scientists merely contracted farmers to provide land or services, or when scientists had consulted farmers about their problems, without involving them in technology design. The influence farmers exerted on scientists' research agendas remained limited. Matters in the international research system were somewhat better, but the record still left considerable room for improvement.

Outside the formal research and development system, different lessons were learnt. In the mid 1980s, disappointed by the performance of the formal system, non-government organisations (NGOs) began working more intensively with farmers at village level, concentrating on LEIA areas. The knowledge and technologies of these farmers became the starting point for informal research and development, using methods such as trial and error and learning by observation. In contrast to the formal system, technologies were to be developed and improved by the people who actually used them. Other farmers, rather than researchers, were seen as the main source of innovations. While criticising the formal system, which was seen as largely irrelevant, indigenous knowledge was revalued, being considered to require a sympathetic external catalyst to become once again the major source of innovations that it had been in bygone ages.

This approach soon showed that it too had its limitations. The improvements brought about through informal research and development were too small to deal with the immense problems afflicting some LEIA systems, which needed more radical change.

*Many experts believe that it will not be possible to feed future world populations while, at the same time, protecting the natural environment, until the full potential of genetic engineering has been realised in world agriculture.* Dr. D. Tribe, Feeding and Greening the World, CABI International, 1994.

Although it could strengthen the self-respect of the rural community, the informal system was too introspective, its practitioners often remaining unaware of the goods and services offered by the formal research and development system. It was also not forward looking enough, being unable to anticipate the risks and opportunities created by changes in the production environment or in farmers' access to resources. The limitations of farmer participatory research in the informal system are summarised in Rhoades and Bebbington (1988).

These perceptions led to attempts to integrate the formal and informal research systems. Meetings organised by the Institute of Development Studies (IDS), Sussex, England in 1988 and by ETC-ILEIA in 1989 brought the NGO and scientific communities together to assess the strengths and weaknesses as well as complementarity. Farmer Participatory Research (Chambers, R. et al., 1989) and Participatory Technology Development (Haverkort, B. et al., 1991) refer to approaches that aim at strengthening farmers' capacities to experiment and innovate. The new approaches were now seen as a complement to station-based research and researcher-managed on-farm trials, not a substitute for them. They allow for a complementary process which links the power and capacities of agricultural science with the priorities and capacities of farmers. Efforts at integration met with some success at the adaptive research level, but formal researchers in applied and strategic research tended to remain aloof. To some extent this was understandable, since these researchers felt they already had their own links with farmers through conventional on-farm research.

As we have seen, the formal system spurned by the NGOs in the mid 1980s had had little impact on LEIA systems, particularly in sub-Saharan Africa. A decade later, that perception must be qualified. Vast problems remain, but for several areas and commodities there are now signs of a modest recovery in LEIA agriculture, driven partly by the availability of new technology. Meanwhile, laboratory-based biotechnology research has come up with an array of new tools with which to tackle the problems faced by resource-poor farmers.

The time is thus ripe for a more profound integration of the informal and formal research and development systems. The two have much to offer one another. Laboratory-based biotechnology research, in particular, stands to benefit from greater emphasis on the client's needs. Conducted at several removes from farmers' fields and requiring a high degree of specialisation, such research runs special risks of developing expensive but unwanted technologies. On their side, researchers in the informal system urgently need access to the new tools and technologies available in the formal system.

Much private-sector industrial research and development demonstrates that specialisation and a strong client orientation are not incompatible. In industry the criterion defining success is clear, the product must reach the market and must sell. The same criterion could guide public-sector agricultural research.

## A contest and a book

This book's protracted period of gestation began in 1992, with an international contest on experiences with rural people's biotechnologies. The idea for this arose from an advisory committee established by the Netherlands Ministry of Development Cooperation, which was at the time launching its own programme on biotechnology research. The aim of the contest, which was organised by the ETC Foundation, was to identify farmers' existing biotechnology practices as a basis for research and development initiatives that would support them.

The contest received some 40 entries. Several competitors described practices hitherto largely unperceived by the research and development community. The winning entry was a 25 page article on "The Indigenous Fermented Foods and Beverages of Sudan". Its author, Hamid A. Dirar, had spent six years documenting the age-old techniques of food fermentation in his country, drawing on the rich knowledge of elderly rural women (Box 1.1).

Most entries to the contest were in the field of food processing, in which indigenous biotechnology is at its most conspicuous and sophisticated, but others covered such areas as plant-based biopesticides, animal health and crop genetic resources. A few contributions concerned biofertilisers and other soil improvement techniques. After the contest the editors decided to explore indigenous knowledge and practices in these fields still further, by inviting a group of experts to write an overview paper on each (1).

These papers, adapted by the editors and preceded by a brief essay written by a farmer, form the basis for Part 1 of this book, devoted to farmers' indigenous biotechnology. This is the starting point for all efforts to improve the lot of the poor in developing countries. Part 2 switches the focus to science-based biotechnology research, which must support farmers' efforts. This part has two chapters, one in which we assess the potential of existing technologies and a second that explores the socio-political context of formal-sector research. Part 3, called Building on Farmers' Practices, consists of a single chapter which sets out a model for integrating the formal and informal research and development systems.

## Questions of balance

In several senses, biotechnology today hangs in the balance.

First, it is poised between the old and the new. Will the traditional biotechnologies, still widely applied around the farmsteads

**Box 1.1**
**Traditional fermentation products in Sudan**

The history of fermented foods in Sudan can be traced back to at least the Meroe dynasty (690 BC–323 AD). Most fermented foods appear to have been developed to ensure food supplies for the family during the long dry season or as survival foods in drought years.

Today, at least 90 fermented foods are made in the country, mainly by women. The Sudanese seem to ferment just about anything edible, even what is barely edible. In addition to the conventional raw materials such as cereals, milk, fish, meat, fruit and honey, unorthodox materials such as bones, hides, hooves, caterpillars, locusts, frogs and cow urine are fermented as delicacies and/or pounded into powders for use as condiments in sauces.

Fermented sorghum products stand out as the most sophisticated, being prepared according to the most complicated recipes. The women make about 30 different fermented foods and drinks from sorghum, which is the traditional staple in rural areas. These products appear unique in several respects compared with other traditional African products. First, about 12 types of sorghum bread are prepared. This is surprising, as Africa is not famous for its breads. Second, several foods and drinks are made from malted sorghum grain, including not only opaque but also clear beers, which are also not common

in Africa. Energy-rich and easily transportable food for travellers is made from sorghum malt: before eating, only water need be added, and the food swells to 3–5 times its original volume.

Much of the three million tonnes of milk produced annually in Sudan is fermented into some kind of dairy product. These vary from lightly fermented by-products of butter to a thick brownish and rancid product (*biruni*) which can be ripened for up to 10 years. Some dairy products are truly indigenous, while others were introduced from the Mediterranean and Middle East about 100 years ago. Many other fermented food products are made from meat, fish and vegetables.

Fermented drinks are equally diverse. In northern Sudan, several types of wine are made out of dates. In southern Sudan, a mead of fermented honey, called *duma*, is a favourite. The *duma*-making process is unique: it is very fast, taking less than 12 hours; organisms which tolerate heat are involved; and the key link in the process is a special starter culture called *iyal-duma* (seeds of *duma*), which immediately triggers fermentation when diluted honey is added. Every family brewing *duma* for sale keeps its starter as a secret, transferred from mother to daughter. It is made from the roots of certain trees through a painstaking enrichment technique, then turned into a paste consisting of an aggregate of a capsulated bacterium and two kinds of yeast.

If washed thoroughly with water and sun-dried after use, *iyal-duma* can be kept for years without losing its potency. This is an advanced biotechnological process carried out at the cottage level.

The women are highly innovative, using their traditional knowledge as a basis for further experimentation. When the Muslim Arabs entered Sudan some 600 years ago, their African wives were faced with a husband who fasted during the holy month of Ramadan and who was hungry and thirsty by sunset. Their response was to invent two new products specially for breaking the fast. The first *wasulu-mur*, a light alcoholic drink containg 31% absorbable sugar. This replenishes glucose in the blood. The second was *abreh*, which is made from thin, transparent sorghum flakes. When mixed with water, the result is a sweet, slippery suspension that slips down the throat without being chewed. *Abreh* and many other thirst-quenching fermented products have lactic acid as an ingredient, and it is possible that this acid has an effect on the physiology of thirst.

Today, *abreh* is still considered the finest sorghum product of Sudan. And the women are as innovative as ever, marketing blue, green, yellow and pink forms of it to keep up with changing consumer preferences.

Source: Dirar (1993).

and in the kitchens of the South, be superseded, as they so largely have been in the developed world, by the ready-made products of the food processing and agricultural input supply industries?

Second, most science-based biotechnology research so far has tended to benefit the high-external-input agriculture of the North. Can this balance be tipped, so that more benefits flow towards the

Woman farmer in Nigeria demonstrating the final stage of a locally discovered cassava fermentation. Cassava is peeled, left to ferment in pots in the sun and then sieved. It is ready for consumption after cooking and final pounding. (*Onyebuchi G.C., who also contributed to the international contest on rural people's biotechnology.*)

resource-poor farmers of the South? The experience of the past does not, as we have seen, augur well in this respect. Yet it is too early to judge. Some institutions are conducting exciting work that holds out great promise that history will not simply repeat itself.

Third, will science-based biotechnology help to protect the environment or merely accelerate its degradation? This is a vast and complex question, but a simple example will illustrate the issue. Molecular markers and genetic engineering now enable scientists to access and transfer to crop plants, genes that confer resistance to a wide range of pests and diseases. This implies a dramatic reduction in the use of chemical pesticides. Yet the very same set of techniques enables private-sector scientists to develop crop varieties resistant to herbicides, encouraging the use of these chemicals.

These three questions are closely related to another issue of balance, one that is perhaps the most difficult of all to resolve: to what extent should the knowledge generated through biotechnical research remain in the public domain, for use in the development of public goods, and to what extent should it be appropriated by the private sector? The recent tendency to allow the patenting of all transgenic varieties for a whole species is widely seen as a most serious threat to the future of advanced research for the benefit of poor people in developing countries.

In a deeper sense, biotechnology today is almost everywhere under intense public scrutiny. Do the possibilities it opens up represent a net benefit to mankind, or an unjustified interference with the natural order that will eventually bring retribution? This question, which arises perhaps most acutely in connection with human medical biotechnology but is also raised by several agricultural applications, is beyond the scope of this book, but it lies behind much of the mistrust felt by those who follow developments in biotechnology.

For some, modern biotechnology has already been 'weighed in the balance and found wanting'. Many NGO staff, in particular, feel it can never benefit their constituency, the resource-poor farmers of developing countries. Yet, as this book will show, there are also grounds for optimism. Farmers themselves possess a wealth of indigenous biotechnological knowledge. The conditions are in place for a synthesis of the formal and informal research systems. A participatory approach will ensure that farmers' needs come first, in the continuing struggle to build a more sustainable, equitable and productive agriculture in developing countries.

## Notes

(1) Only the first four fields are represented in this book. We were unable to find an expert to cover soil improvement.

## References

Brown, L., Kane, H. and Roodman, D.M. 1994, Vital signs 1994: *The Trends that are Shaping our Future*, Worldwatch Institute, Washington DC, USA

Chambers, R. and Ghildyal, B.P. 1985, 'Agricultural research for resource-poor farmers: The farmer first and last model', Agricultural Administration 20: 1–30

Chambers, R., Pacey, A. and Thrupp, L.A. (eds), 1989, *Farmers First: Farmer Innovation and Agricultural Research*, Intermediate Technology Publications, London

Dirar, H.A. 1993, 'The indigenous fermented foods and beverages of Sudan', paper submitted to the Contest on Rural People's Biotechnology, ETC Foundation, Leusden, Netherlands

Haverkort, B. van der Kamp, J. and Waters-Bayer, A. 1991, *Joining Farmers' Experiments: Experiences in Participatory Technology Development*, Intermediate Technology Publications, London

McCorkle, C.M. (ed), 1989, *The Social Sciences in International Agricultural Research: Lessons from the CRSPs*, Lynne Rienner Publishers, Boulder (Colorado) and London

Pinstrup Andersen, P. and Pandya-Lorch, R. 1994, *Alleviating Poverty, Intensifying Agriculture, and Effectively Managing Natural Resources*, International Food Policy Research Institute, Washington DC, USA

Reijntjes C., Haverkort, B. and Waters-Bayer, A. 1992, *Farming for the Future: An Introduction to Low-external-input and Sustainable Agriculture*, Macmillan, London, UK

Rhoades, R. and Bebbington, A. 1988, 'Farmers who experiment: An untapped resource for agricultural research and development', Paper presented at the International Congress on Plant Physiology, New Delhi, India

*Biotechnology research, by and large, does not focus on the needs or interests of poor farmers in marginal areas of the world.* FAO Expert Consultation on Harvesting Nature's Diversity, 1993.

Tribe, D. 1994, *Feeding and Greening the World*, CABI International, Wallingford, Oxford, UK

Uphoff, N. 1992, *Local Institutions and Participation for Sustainable Development*, Gatekeeper Series, International Institute for Environment and Development, London

Wolf, E.C. 1986, 'Beyond the green revolution: New approaches for third world agriculture', World Watch Paper No. 73. Worldwatch Institute, Washington DC, USA

# Part 1

# Indigenous Biotechnology

# 2 A Farmer's View

*Michael N. Kibue* (1)

## My Farm

My heart is in agriculture and each day of my life starts with me on my farm.

Our Kikuyu community is traditionally agricultural, occupying the slopes of Mount Kenya and the Aberdares, which have well-

My heart is in agriculture and each day of my life starts with me on my farm. (*Kees Manintveld*).

drained volcanic soils and moderate rainfall. A typical Kikuyu landscape in our area is marked by freshness and greenness. The soils are red-brown on the hillsides and grey-black in the valleys. Patches of indigenous forest are like islands of deeper greenery, while cultivated plots separated by rows of shrubs and trees give the hillsides the beautiful look of productive land that is well cared for.

The Kikuyu used to practise mixed farming in which the animals; mostly cattle, goats, sheep and donkeys, were part of the farm. Our main crops were maize, bean, sorghum, millet, cowpea, yam, banana, sugar cane and indigenous vegetables. A family cultivated several small plots over a wide area. This enabled them to grow crops on different soils, using hillside and valley gardens, which were suited for different crops. It also guarded against losses due to pests and diseases. The wealth of our community was measured in terms of production: the more animals and granaries, the richer the community.

This traditional system has drastically altered in recent years. Many farmers have switched over to cash crops such as coffee, tea and pyrethrum. On these and other crops they are using increasing amounts of chemicals. Some indigenous food crops – yam, sorghum, millet – have all but disappeared from our area. Animal production systems have also intensified, with the introduction of dairying and pig husbandry.

Yet today's small-scale farmer is at a crossroads. He/she is unable to decide whether to proceed with modern agricultural practices or with the traditional ones. Modern technology was supposed to make our work easier and to help us achieve maximum returns to our labour. But it has also caused extensive damage to our environment. Worse, we cannot afford to continue practising modern agriculture due to the high cost of inputs, which are rising in price day by day.

Our farmers do not like using chemical fertilisers because it makes their soils hard and infertile. The long-term infertility is caused by the drain of micro-nutrients, brought about by mining of the soils. Residues of harmful pesticides have increased in our water and food supplies, causing health problems. Using chemical inputs has not only impoverished the farmer but also threatened to exterminate his co-workers, the tiny micro-organisms that have tirelessly sustained agricultural production over generations.

I own a two-acre farm in the lower part of Kiambu District of Central Province, where I practise mixed crop and livestock production and conduct experiments. My previous attempts at modern agriculture having failed, I started to search for a more sustainable and productive system that would reduce our dependency on expensive fertilisers and chemicals yet increase our output and incomes. My approach is to enrich our traditional practices with modern science, capturing the best of both worlds.

# The role of biotechnology

## Traditional practices

Several traditional practices in our community reveal indigenous biotechnology at work.

Soil fertility and the protection of the natural resource base were secured through the natural processes of shifting cultivation, bush fallowing and composting. In our community, trees were valued greatly. They were classified as "crop-friendly" or "crop-unfriendly". Crop-friendly trees such as *Commiphora eminii,* locally known as *mukungugu*, were interplanted with yams.

Pests were not a major problem to agriculture under our traditional system. Our farmers managed to keep infestations at low levels by maintaining biodiversity through the intensive mixed cropping system, known as *githoboco*. The word means "random mixed seeding in the field". The density and spacing used in this system were based on the experience of the woman farmer. Maize, millet, sorghum, bean, sweet potato, yam, banana, cassava and sugar cane were all intercropped, and this reduced pest attack. Ash was used to control the small worms that attack seedlings and to deter attack by small black ants and termites on banana and yam.

Our forefathers recognised that the plant was a complex chemical factory. For many generations, these natural factories made compounds which were used in the treatment of various human and animal diseases. For example, *Brucea antidysenteria (mukuriuahungu)* and *Warbugia ugandesis (muthiga)* were used to treat abdominal ulcers, weak joints, diarrhoea and malaria. The mode of application was to extract the active substances from the roots, leaves or bark of the plant by boiling with water, then mix the concoction with bone soup or honey syrup. The mixture was allowed to ferment before being administered to the patient.

Similar practices existed for animal diseases. For example, *Heliotropium* spp *(muramata)* was mixed with other plants, ground and mixed with water for use as a drench to treat East Coast Fever after the swollen glands of affected animals had been branded with a hot iron bar.

Genetic conservation was once part of the farming system. Each farmer was both a conserver and a breeder of plant genetic resources. Women farmers ensured that there were sufficient seeds for planting, selecting the best seeds from last season's harvest. For example, for maize a healthy comb with no empty spaces was chosen. The seeds were removed only from the middle part of the comb, so their high quality was ensured. For legumes, healthy seeds were selected according to size, colour and taste.

Farmers maintained genetic diversity by borrowing or exchanging seeds. Even today, our farmers are very generous with their seeds. Animal breeding was the responsibility of the man of the family. He bought or borrowed diverse breeds to cross with his own stock. The size, colour and sure-footedness of the animal were the basic criteria for selection.

Food processing was traditionally the speciality of the woman farmer. Natural fermentation processes were widely used to prepare foods and beverages. Examples are sour porridge (*uchuru*), liquor (*njohi*), fermented meat (*rukuri)* and sour milk (*iria imata*). The major fermented food prepared was the popular *uchuru* , made by pounding cereal grains such as maize, millet or sorghum and mixing them with a little water in a mortar. The uncrushed residue was filtered off and the thick filtrate diluted, warmed, then placed in gourds to ferment. After three days the sour porridge was served. A similar procedure was used to prepare *njohi* from sugar cane.

## Building on tradition

We must build on these traditions as we undertake to develop biotechnological innovations for the small-scale farmer. Our efforts should be directed towards modifying and improving traditional biotechnologies and to strengthening or restoring farmers' rights. To avoid antagonising nature, innovations should be based on natural rather than synthetic mechanisms.

With these considerations in mind, I propose the following sustainable biotechnologies for today's small-scale farmers.

### Soil fertility

The farmers' traditional methods of replenishing the soil, fallowing and composting can be accelerated through the following approaches:

- Activated composting. Accelerated microbial degradation of organic matter through solid-state fermentation can be developed for use on the farm. With this technology, the farmer would overcome the problems of availability of biomass and the cumbersome procedure of traditional composting. He/she would also make maximum use of available organic resources.
- Improved simultaneous fallowing. The best alternative to shifting cultivation is improved simultaneous fallowing through multiple cropping systems involving a wide range of leguminous plants, trees, hedges and tuber plants grown with food crops in a deliberate configuration. We need to establish the best plant combinations and sequences. Some farmers in my area are already developing new approaches in this field.
- Improved nitrogen fixation. The ability of leguminous plants to fix atmospheric nitrogen is now well understood. Biotechnologists could assist small-scale farmers in selecting the best combinations of leguminous and grain crops to make use of this ability. Microbial inoculants to accelerate N–fixation should be made more stable and practical for use by the farmer.

## Crop, animal and human health

There is great potential for herbal medicine today. Unlike in the 1960s and 1970s, when our traditional medicine was regarded as witchcraft, many people, including myself, today value traditional medicine. Biotechnology can help our traditional doctors develop better preparations and prescription methods.

Extracts from various plants have already been used to make biopesticides. Others could be developed. Thorough research will be needed to ensure that the products are both safe and effective.

## Genetic resources

There is an urgent need to stop genetic erosion and guarantee stable agricultural production in the future by conserving the genetic resources of food crops. This will only be possible if the farmers' right to be the custodian of plant genetic resources is restored.

Many farmers blame hybrid seeds for their poor harvests and have a great need to produce their own genetic resources. Simple seed production and breeding practices should be developed and promoted among farmers. Practices would include tissue culture, vegetative culture and natural cross-breeding. Tissue culture of potatoes can easily be carried out in laboratories managed by farmers.

## Renewable energy

Biotechnology can play an important role in the production of fuels from renewable sources. Methane, ethanol and methanol can all be produced through energy-intensive biochemical reactions. The major inputs are organic residues.

However, production of these fuels at farm level is difficult. Methane can only be produced when there is plenty of animal manure. Production is practical only if integrated with other essential functions. For example, methane can be a by-product of the activated composting process.

## Food processing

I am happy to mention that, in this field, our community has agreed to return to one of its traditional and best loved biotechnologies. Our mothers have vowed to serve our fermented porridge, *uchuru*, at all social gatherings, replacing the expensive drinks previously served. Although this noble offer has been accepted and implemented since early 1994, the practice is once again threatened because the use of plastic containers for fermentation makes the *uchuru* bitter. The traditional *nyanja* (gourd) used to ferment *uchuru* makes it very sweet. So, lack of proper equipment may thwart the re-introduction of our good old traditions.

Better fermentation procedures for our food products could be developed. The enzymes used should be made more efficient, so that they can be deployed to produce a wider range of food products. In particular, solid-state fermentation could be better controlled, helping the farmer diversify food production for the family. For example, it could be used to make animal feed from green biomass harvested during the rainy season. The fermented feed could be stored as blocks for use in the dry season. However, the bulkiness of most raw materials used in the fermentation process can make the results uneven.

There is an urgent need to develop appropriate biotechnology tools and equipment for use at the small-scale farm level. These would include shredding and pounding equipment, fermentation vessels, stirring and separation equipment, packaging plant, and so on. Luckily we have some indigenous fermentation tools which can serve as a basis for designing improvements. Examples include the pestle and mortar, which are made of wood, the calabash, and several other utensils derived from the fruit of plants or made of clay. Each tool had specific applications. I am greatly concerned lest these valuable indigenous tools should shortly become extinct and so lost to future generations.

## Conclusion

There is an urgent need to reverse the negative trends we see in agriculture. Sustainable forms of production, sympathetic to farmers' social and economic circumstances and to the natural resources available, need to be developed. Higher production will prove sustainable only if the natural resource base is secure.

Small-scale farmers should be seen as a resource, not a liability. Their creativity and capacity to innovate is the basis for achieving increases in production. Their indigenous knowledge, gained throughout generations, should be inventoried and harmonised with modern science to evolve a diversified technology base. Researchers must accept that they need to learn from the farmer and to build on farmers' innovations. But the farmer too must fill gaps in his or her knowledge and learn new technologies that will underpin sustainable gains in production. Indigenous biotechnologies, like modern scientific ones, are dynamic. We should not seek to return to medieval times by attempting to use technologies that are outdated and inadequate to meet the challenges of today.

The responsibility for genetic conservation, the guarantee of stable future production, should be vested in the farmer. The unfavourable marketing and pricing policies that marginalise food crops in favour of cash crops will have to be changed. Unfortunately for poor farmers, the present rules of world trade are very unfair. Nevertheless, farmers will have to take risks and innovate. Only then do they stand a chance of bettering their lives. The challenge here is one of taking action now, not just waiting for policy changes.

*We do not say this is the only way to do it. No, we say for us at this moment this is the right thing to do. We see other farmers in different places have different results. That is nice, it is not a problem. In ten years time, everything may have changed and our practices may no longer work. But we will find new ways by working together. We must not convince others, we must convince ourselves. That is the only way.*    Paysans sans Frontières, by Doudon Sow and Remi Shiffeleers, in ILEIA Newsletter, 1995, 11(4): p. 9.

On a small farm, thinking and acting are united in one person, who must decide whether or not to incorporate ideas from the outside. Although we can join networks and NGOs and get advice from our friends, in the end we are on our own in making these investment and management decisions. To me, a farmer must always convince him or herself that an idea works, before trying to convince others. And what works in one place or time may not work in others, so life is risky. The quotation in the margin, which is by some fellow farmers, reflects my feelings exactly.

## Notes

(1)  This an abridged version of one of the papers contributed to the Contest on Rural People's Biotechnology.

# 3 Animal Health

*Evelyn Mathias and Constance M. McCorkle*

## Introduction

Over centuries, stockraisers, both farmers and herders, have developed their own ways of keeping their animals healthy and productive. They treat and prevent livestock diseases using sometimes age-old home remedies, surgical and manipulative techniques, husbandry strategies and associated magico-religious practices. Taken together, these indigenous and/or local animal health care beliefs and practices constitute what is now known as ethnoveterinary medicine (McCorkle, 1986; McCorkle, 1995).

Many ethnoveterinary practices can be classified as simple biotechnologies according to the definition used in this book (see p. 1). Stockraisers have long manipulated micro-organisms and the cells and tissues of plant and animal species (including those of infected livestock) to produce crude vaccines or to otherwise protect animals against disease, to yield useful veterinary drugs, and generally to promote the health and productivity of their herds and flocks. The production of traditional vaccines and other drugs can be characterised as *in vitro* biotechnologies, while the management of disease-causing organisms within the host animal's body or the manipulation of the patients' cells and tissues qualify as *in vivo* biotechnologies.

This chapter provides an overview of these ethnoveterinary techniques and, where possible, offers some assessment of their validity from a Western scientific point of view. It then discusses the cultural and socio-economic context in which ethnoveterinary knowledge is generated and applied. Finally, it assesses the importance of ethnoveterinary medicine for stockraisers worldwide and outlines future research needs in this area. Throughout, the authors draw upon their own experience in ethnoveterinary research and development in Africa, Asia and Latin America, plus a wide range of published and unpublished reports and accounts (1).

## The biotechnological context

This section discusses ethnoveterinary practices that meet the definition of biotechnology used in this book. Excluded are most husbandry and nutritional strategies normally dealt with as part of ethnoveterinary medicine. Also excluded are breeding techniques based on selection rather than on the manipulation of cells and tissues (for example artificial insemination, embryo transplant), except where these affect disease resistance and control.

## *In vitro* techniques

### Vaccine preparation

Stockraisers worldwide have developed techniques for immunising their animals against a number of viral, bacterial and protozoal diseases. The most familiar technique is inoculation with a vaccine. Among the viral diseases for which crude vaccines have been developed are sheep and camel pox, foot-and-mouth disease (FMD), contagious ecthyma, rinderpest and (deMaar, 1992) warts. The bacterial diseases treated in this way include blackleg, brucellosis, contagious bovine and caprine pleuropneumonia (CBPP, CCPP) and (Baekbo and Nielsen, 1992) swine diarrhoea. Box 3.1 describes a traditional vaccination technique practised in West Africa.

Immunisation is normally done for one of two reasons: to prevent diseases known to have high mortality rates; or to control diseases that, while usually non-life-threatening, can greatly reduce herd or flock productivity. Control is typically achieved by vaccinating the herd or flock so as to induce a mild case of a disease at an opportune time of the year. In this way, the disease will not strike when animals are in poor condition (due, for example, to feed scarcity or to pregnancy and parturition) or while they are being trekked.

---

**Box 3.1**
**A home-made vaccine**

Fulani pastoralists in northern Nigeria prepare and apply a home-made vaccine to combat CBPP. They remove the lungs of a diseased animal, cut the lung tissue into small pieces and ferment it in milk for two or three days. Then they insert the pieces under the skin of the forehead of their cattle. The incisions are sealed with mud. After three days they are re-opened and the lung tissue is removed. The wound is washed and the surrounding area is cauterised.

This technique almost always causes painful local swelling and often leads to more extensive swelling due to bacteriological contamination. The wound may suppurate, and the animal may become feverish. Early removal of the lung tissue and cauterisation are clearly important for preventing more severe effects. Even so, the wound may take months to heal.

The pastoralists claim this practice protects their animals for at least a year. Its widespread use despite its unpleasant side-effects suggests they are right.

Source: Leeflang, P., Veterinary Health Service, Ministry of Agriculture, The Hague, Netherlands.

To prepare their home-made vaccines, stockraisers use biological materials collected from livestock that have contracted the disease in question. Such materials include faeces, blood, saliva, pus and lesion or organ tissues.

These materials may be applied fresh and unaltered. For example, Indian farmers vaccinate against CCPP by taking blood or a scrap of tissue from the ear of an infected goat and then pricking the blood into the ears of goats or temporarily inserting the tissue into a notch in the ears made for this purpose. Another example from India is a traditional FMD vaccination for cattle that is accomplished by wadding a cloth into the mouth of an infected animal and then wiping the saliva-soaked cloth across the lips of other cattle (personal communication, 1994).

Alternatively, the immunising materials may be modified through a variety of processing or compounding methods that serve one or more purposes. For example, vaccine production typically entails reducing the virulence of the materials used. Otherwise, they might kill rather than immunise the patient. In Western medical parlance, this process is called attenuation. Among herders of the African savannas, at least four attenuation techniques corresponding to recognised scientific methods have been identified; desiccation, petrification, dilution and fermentation. Some of these processes also enable stockraisers to store the infective materials for an extended period.

To illustrate, herders in India and Arabia collect and dry camel-pox crusts/scabs from naturally infected animals; then, when pox vaccine is needed, they pulverise the dried material, mix it with milk, dip a thorn or needle into the mixture and prick the camel's lips with it. Fulani pastoralists of the African savannas prepare a rinderpest vaccine by aqueous maceration of the lung tissue of cattle that have succumbed to the disease. Somali pastoralists similarly prepare a CBPP vaccine from minced lung tissue washed in salt water. For rinderpest, they compound urine, faeces and milk from an animal with a mild case of the disease; they say this vaccine will remain active for several months if it is properly stored. However, a set of samples sent for examination could not be proved infective (Mares, 1954).

Schillhorn van Veen (1996) considers indigenous vaccines for poxes, FMD and ecthyma effective. He views the efficacy of other vaccinations as dubious (blackleg, rinderpest) or variable (CBPP, CCPP, brucellosis). For example, across many years of field veterinary practice, it has been observed that the Indian CCPP immunisation described above is effective in about half the animals treated (personal communication, 1994). Indeed, the efficacy of any crude vaccine can vary greatly, depending on several often interacting factors – such as the virulence of the immunisation materials, the dosage given, animal weight and condition, and mode of application.

Fulani pastoralists in northern Nigeria protect healthy cattle by inserting a home-made vaccine under the skin of the forehead of their cattle.
(*Paul Leeflang*).

## Drug preparation

People everywhere make a wide variety of medicines to treat or condition their livestock. Biological materials from the animal kingdom may figure in such drugs, which may be made from substances as diverse as bone, horn, dung, urine, milk, honey, termite earth, snake skin, animal fats and crushed or burned body parts of wild animals. Minerals of many sorts are also commonly used, including limestone, potash, clays, sodium chloride and other salts, sodium carbonate (baking soda), iron and iodine-rich substances, and phosphorus from various sources. But by far the majority of traditional drugs rely primarily on plant cells and tissues, whether from food or non-food crops or from a host of trees and shrubs.

Stockraisers may collect medicinal plant materials only at certain times of the day or the year, sometimes coupling this task with prayers and magico-religious acts and ceremonies. The whole plant may be required, or only selected parts, such as leaves, bark, rhizome, root or tuber, root/tuber bark, fruit, flower, bud, pistils, stamen, stem, seed, sap/latex/resin, and so on. Again, the plant material may be fed to patients fresh, or it may undergo processing first. Often, it is air-dried for storage. Further processing typically depends on the route of administration and, presumably, on the active ingredient(s) to be extracted. Although stockraisers may not name or recognise these ingredients *per se*, certainly they appear to have observed through trial and error that some preparations of a plant are more effective than others.

Most ethnoveterinary remedies are compounded from several different plants. To take but one among untold thousands of examples worldwide, farmers in West Java combine tamarind fruit pulp, tumeric rhizome, and *Areca catechu* sap to produce an anti-diarrhoeal for small ruminants (Ma'sum, 1991). Only rarely does a given ethnopharmacopoeia employ mainly single-plant remedies. An exception is that of Kenya's Samburu pastoralists, who overwhelmingly employ monobotanical preparations. This is suggestive of a profound indigenous knowledge of empirical bioassay among these people (McCorkle, 1994a).

Drug preparations span infusions (both hot and cold), decoctions, powders, drops, ashes, pastes and boluses (sometimes prepared by chewing the plant materials), lotions, emulsions, ointments, fumes, smoke and vapours. There appear to be some cultural preferences or patterns in ways of preparing plant cells and tissues. In Java, for instance, mechanical procedures like chopping, crushing, grinding and pounding predominate (Mathias-Mundy and Murdiati, 1991). Of some 70 plant preparations used by Javanese villagers to combat small-ruminant ills, only seven called for heating or burning (Wahyuni et al., 1992). In contrast, Quechua agropastoralists in the high Andes of Peru appear to favour cold infusions and, to a lesser extent, smoke cures. Still other cultures give pride of place to non-botanical treatments. For example, Turkana pastoralists in Kenya swear by branding and cauterisation to prevent and treat most livestock ailments.

Out of the estimated 10% (25 000 to 75 000) of higher plant species that have been used in traditional medicine, only some 1% are acknowledged through scientific studies to have real therapeutic value when used in extract form by humans (Farnsworth, 1983). Although comparatively little research has so far been devoted to ethnobotanicals for livestock, many of the same plants also figure in human ethnomedicine and have been studied in that context (e.g. Anjaria, 1996). Such studies, together with the few specifically veterinary studies available, indicate that many species do in fact display pharmacological activity against microorganisms such as viruses and bacteria (see the reference list in Matzigkeit, 1990; or Roepke, 1996). Research on samples of African ethnoveterinary botanicals suggests that as many as 30% may be effective against the livestock diseases they are used to treat. A similar figure seems reasonable for the Andes. Such estimates parallel the 20% figure commonly cited for human ethnopharmacology, although in some pharmocopoeia this figure can reach 80% (Trotter, 1988).

Nevertheless, scientific validation of the action of traditional botanicals has often been problematic. For a variety of methodological reasons, formal pharmacological and clinical studies have not always been accurate in their assessment of the activity, efficacy or safety of these remedies. This is especially true of polyprescriptions, but is also the case for monobotanical preparations. Studies sometimes fail to take into account the location from which the plants are normally collected; yet soil types can influence phytochemical balances and hence pharmacological effect. Likewise for the developmental stage at which the plants are collected; young versus mature, in or out of flower, and so on. Further complicating factors are the additive, synergistic or antagonistic pharmacological effects that can occur when plant substances are compounded. Finally, validation studies may not faithfully reproduce local preparation procedures and the quantities of materials used, yet these too can influence efficacy. Even something as simple as chopping or shredding the plant materials with a metal instrument instead of a traditional wooden or bamboo tool might make for chemical differences.

Research on a traditional Thai anthelmintic for humans illustrates some of these problems. The Thai make an effective dewormer by pounding *Diospyros mollis* and mixing it with coconut juice. But when pharmacologists tried to mass-produce this remedy using an electric blender and leaving the mixture to stand overnight, unanticipated chemical reactions generated toxins that caused blindness among patients (CUSRI/CESO, 1988). To take another example, in the 1940s US and European pharmaceutical firms tested a tree sap used by the Amazonian Indians to treat fungal infections. Assayed only in dried form, the sap was found to have no active principles. But 40 years later, when it was re-tested in the fresh form in which these Amerinds were, fortunately, still using it, the sap yielded three new compounds, two of which proved to be antifungals (Plotkin, 1988).

A parallel but converse ethnoveterinary example comes from on-station experiments to validate an indigenous Javanese treatment for endoparasites of sheep. To control and standardise the dosage of the key ingredient involved (sap from immature papaya fruit, scientists collected and administered only the sap itself) in contrast to the farmers' practice of feeding the whole fruit. In a matter of hours after treatment, 80% of the sheep in the high-dosage experimental group died of acute poisoning (personal communication, 1994).

The larger lesson in the foregoing examples is twofold. First, for accurate scientific validation of indigenous practices, it is important to duplicate them as closely as possible. This may prove difficult for Western scientists to accept, especially when these practices are highly unorthodox. An example is an eyewash prepared by human mastication of plant materials followed by spitting into the animal's eye, a practice observed in Kenya (McCorkle, 1994a). Second, it may be necessary to devise innovative research designs, rather than mindlessly cleaving to conventional "canned" methodologies.

## *In vivo* techniques

Ethnoveterinary techniques may also involve the management of micro-organisms *in vivo*, that is within the bodies of host animals. *In vivo* techniques can be divided into three types according to whether the intent is to prevent/control problems (prophylactic), to cure ills (therapeutic), or to enhance productivity (promotive). These three intents can in turn be expressed in terms of two categories of action: management of micro-organisms, primarily via the application of traditional immunisations and drugs; and manipulation of the patient's tissues, both directly and indirectly.

### Management of micro-organisms

Immunisation probably constitutes the most dramatic example of indigenous prophylactic techniques involving the management of micro-organisms *in vivo*. The intent is to confer immunity, of variable duration, through one of two mechanisms: by introducing into the patient an attenuated or limited quantity of the infective agent of a disease, so that the host animal will build up antibodies to it; or by introducing antibodies from another animal.

Vaccines prepared *in vitro* are usually applied *in vivo* in one of three ways: orally, by drenching or feeding; by simple inoculation, using a thorn or needle; or by surgical implantation (see Box 3.1).

Besides such direct forms of vaccination, an indirect method of *in vivo* immunisation is known to some stockraising peoples. This consists of deliberately exposing one's flock or herd to infected animals. For example, Indian farmers traditionally placed poultry suffering from coccidiosis or fowl pox (a protozoal and a viral disease, respectively) in with young, uninfected birds so that the latter would also acquire the disease and thus some natural immunity to it (personal communication, 1994). Of course, there is a danger in such practices, in that the virulence and dosage of the immunisation cannot be controlled; thus a number of animals may well be lost. More dramatic and quite successful is the Fulani strategy of driving their cattle downwind from a herd with FMD. They know that their animals will thereby contract a mild case of the disease, which confers immunity. The wisdom behind such strategies is impressive. In the case of FMD, for instance, it took until 1967 for Western veterinary scientists to determine that the causative virus could be transmitted aerially across long distances.

One further example of an indigenous immunisation technique is the practice of some stockraisers of ensuring that a newborn animal immediately suckles its dam's colostrum. Colostrum contains antibodies that are thus passed to the neonate, conferring immunity to a spectrum of diseases.

For therapeutic purposes, people have long managed the micro- and other disease-causing organisms such as endoparasites that invade their livestock. They do so by introducing their ethnomedicines into the patient's body, most often in one of two ways: topically, that is on the skin surface, and orally. Less commonly used are nasal, ocular, vaginal, anal and aerial routes of administration.

Wound care provides a good illustration of topical treatments. Stockraisers worldwide often wash wounds with a home-made disinfectant. A wide array of plant- and mineral-based ethnomedicines serve as antiseptics. Additionally or alternatively, stockraisers may debride (cut away infected tissue) or cauterise the wound site. After application of a traditional medicine, the wound may be dressed with protective materials such as kitchen ash, resin or wax; or it may be sutured. From the viewpoint of Western veterinary medicine, such procedures are effective. Cleaning, debriding and cauterising reduce bacteria counts and thus the danger of infection. Suturing the wound or dressing it with relatively sterile materials provides a mechanical barrier to the re-entry of bacteria.

Perhaps the technique most widely used by stockraisers to introduce drugs into the animal's system is drenching, meaning here the force-feeding of medicine in liquid form. This may be accomplished using such items as a soda-pop bottle, a hollow piece of bamboo or a gourd. Performed correctly and with safe materials (thin glass containers are not a good idea), this is essentially the same technique as that used worldwide for administering commercial livestock drugs in liquid form.

Examples of other, less common, routes of administration include: in Southern Africa, injections of coconut water under the skin as a rehydration therapy (Roepke, 1996); throughout most of the world, numerous kinds of eye medicine are administered intra-ocularly (as drops using a hollow straw, a thin bamboo or an eyedropper; as washes (IIRR, 1994); or as "injections" that are spat into the eye). Drugs may also be given intra-nasally, via instillation in solid or liquid form and via inhalation of smoke or steam from burning or boiling the materia medica. The latter treatments are documented in Africa, Asia and Latin America, typically for respiratory ailments. They are often administered by covering the patient's head with a blanket and then holding the head over a small pile of slowly burning medicinal leaves, for example eucalyptus, or over a steaming decoction. A parallel in Western human medicine is the household use of vaporisers.

The ethnoveterinary use of vaginal and anal suppositories has also been attested. A fairly unusual mode of administration is to hang up strong-smelling bouquets in livestock quarters. This practice has been reported in Asia (IIRR, 1994), Africa and Europe. French shepherds, for example, hang bouquets of certain plant and, less frequently, animal species in their sheepfolds, both to prevent and to treat diseases with cutaneous signs, such as ecthyma, ringworm and sheep pox (Brisebarre, 1996). The shepherds say the bouquets operate mainly via aerial diffusion of substances associated with the plants' odour. In fact, out of 26 plant species used in such medicinal bouquets, 16 are or were used externally in veterinary or human dermatology.

Reports of promotive *in vivo* manipulation of organisms are rare in the ethnoveterinary literature. However, one example from Asia concerns the correction of pH imbalance in the rumen by feeding special feeds or drinks. Besides grains and fodders, these include clays, minerals and even soda-pop. Such items can have alkalising or, conversely, acidifying effects. Western veterinary medicine confirms that diet can directly affect stomach pH, and hence the types and numbers of bacteria that flourish in the rumen (McDonald et al., 1973). A related technique used throughout the world by stockraisers and Western-trained vets alike is to feed the cud of a healthy cow to a herd mate that has chronic indigestion or that has just completed antibiotic therapy, so as to rehabilitate the latter's rumen flora and fauna. For these conditions, and for certain diarrhoeas, an ailing horse may be fed the faeces of a healthy one.

The efficacy of nearly any *in vivo* treatment and its associated mode of application varies according to factors such as the disease involved, the type and state of materials used and the condition of the patient. Some treatments act on the disease-causing organisms directly, while others work indirectly to improve the general health of the animal and hence its ability to resist or combat offending organisms. But for still others, such as medicinal bouquets, there is as yet no scientific explanation. While bouquets may serve merely as placebos to calm the worried stockraiser, it is also conceivable that their scent may help to repel pests bearing disease-causing organisms.

A frequently cited benefit of traditional treatments is that, in addition to being generally cheaper and more readily available, they are often safer than the potent synthetic drugs marketed by chemical companies. Some of the arguments for this statement are as follows. Traditional medicines tend to be less concentrated and therefore less likely to be taken in poisonous quantities, as in the papaya fruit example given above. Because traditional medicines are often more complex than synthetic drugs, there may be synergisms or antagonisms that offset negative side-effects. Often, too, traditional drugs are less biostable and biocumulative – that is, they break down more quickly and are less likely to build up in the body, or the environment, in harmful amounts. Indeed, the very fact of a traditional medicine's longstanding use throughout centuries or even millennia testifies to its relative safety (Etkin, 1990; Fugh-Berman, cited in Bodeker, 1994; WHO, 1991).

It is nevertheless possible to overdose with strong local prescriptions, just as with commercial drugs. For example, the first author encountered a case in which a Javanese farmer drenched a ewe with 600 ml of a herbal mixture and divided another 600 ml bottle between the ewe's two lambs. One lamb died immediately, and both the ewe and the other lamb showed signs of poisoning. In addition, there is always the risk of improper or unskilled application of treatments. A danger in drenching, for instance, is that the liquid can be aspirated into the lungs. But in skilled hands this is no more likely to happen in traditional than in modern veterinary medicine. With the exception of medicinal bouquets, all the routes of administration employed in ethnoveterinary medicine have long been used in Western scientific medicine too.

## Manipulation of tissues

This can be accomplished indirectly or directly. Indirect manipulation includes preventive measures such as supplementing the diet to condition animals, or selectively breeding for disease tolerance or resistance. Direct manipulation is involved in techniques such as cauterisation, branding, bloodletting, and all forms of surgery, including many obstetric procedures.

For prophylactic purposes, stockraising peoples commonly provide their animals with supplementary minerals, edible earths, special feeds or tonics. At the very least, they do so during certain seasons or for special classes of stock such as pregnant or lactating females, studs, or traction and pack animals. They know that good nutrition helps ward off disease, increasing production and performance. Of course, this is also a tenet of Western veterinary medicine.

Selective breeding constitutes another mechanism for disease prevention and control. For example, Fulani pastoralists purchase local bulls to enhance their herds' adaptation to local conditions, even if the new bulls are of a different breed from the rest of the herd. Throughout Africa, such strategies have produced unique disease-tolerant or -resistant breeds of cattle, sheep and goats (Schillhorn van Veen, 1996).

As with some of the other traditional techniques described here, selective breeding may not always represent a conscious attempt to manage genetic resources for disease prevention. Often, it is done for other reasons; to achieve a preferred coat colour, fibre type, build, horn conformation, and so on. Magico-religious beliefs may sometimes be involved. For example, the Bori cultists of Nigeria keep certain varieties of chicken that show neurological signs resembling spirit possession in humans. These varieties are used to invoke the birds' associated Bori spirits to assist in human healing ceremonies. Some Nigerian farmers claim that these birds also survive Newcastle disease better than other local chickens. But probably cult members keep and breed these varieties primarily for religious rather than animal health reasons (Ibrahim and Abdu, 1996).

One widespread routine and multi-purpose ethnoveterinary technique that entails manipulating the host's tissues for thera-peutic purposes is cauterisation. It consists of the destruction of tissue (and of any offending micro-organisms therein) by intense heat, applied through a red-hot instrument, a caustic agent, or fire. Cauterisation instruments include hot iron rods and red-hot stones or potsherds. For some conditions, however, stockraisers may sprinkle a flammable substance on the diseased parts of the patient's body and then briefly ignite it. To treat footrot, for example, French shepherds use gunpowder in this fashion, while smallholders in northern Brazil apply gasoline.

The literature cites a wide range of diseases traditionally treat-ed by cauterisation: anthrax, trypanosomiasis, rickettsiosis, epilep-sy, botulism, scabies, bloat, diarrhoeas, toothaches, fevers, diges-tive and hoof ailments, snakebite, and muscle pains to name but a few. In Western veterinary medicine too, cauterisation is a com-mon practice. It is especially used in persistent inflammations of tendons and joints, where it induces a localized hyperaemia, which stimulates the reabsorption of chronically inflamed tissues (Berge and Westhues, 1969). It can also be beneficial for lame-ness, sprains and muscle strains, as well as for suppurating wounds or those caused by castration, docking or de-horning. The value of cauterisation for treating generalised infectious dis-eases, however, is questionable, since it may depress antibody production through the release of histamines (Marx, 1984).

Another widespread ethnoveterinary treatment is bloodletting. Indications cited in the literature range from, curing laziness, through, improving the general condition, to a wide variety of infectious diseases. Until fairly recently, bloodletting was an accepted part of European veterinary practice, especially for bovine and equine medicine.

Stockraisers also perform other surgical operations, such as lancing boils and abscesses, excising tumours, and cutting away necrotic tissue. Properly and hygienically performed, these operations all make good sense from a Western as well as an ethnoscientific point of view. However, some traditional surgical interventions may be ineffective or potentially harmful, serving no scientifically evident health purpose while posing the risk of infection, as in any open surgery. An example is the practice, found among some South American agropastoralists and African pastoralists, of scoring or cutting off parts of an ailing animal's ear so that it bleeds. The illness, or the toxins or spirits believed to produce it, are thought to be expelled with the blood.

While accounts of surgical skills abound in the ethnoveterinary literature, descriptions of direct tissue manipulation for promotive purposes are rare, if one discounts castration. The latter is a classic example of manipulation (usually, removal of the testes) to enhance growth and tameness, mainly in meat and work animals. In an effort to reap both these productivity features while also retaining greater herd fertility, Quechua agropastoralists sometimes remove only one testicle from their pack llama.

A fertility-enhancing manipulation practised by Twareg herders of the African Sahel is the removal of a persistent corpus luteum from a camel's ovary by manual extraction through the rectum (Nicolaisen, 1963, cited in Köhler-Rollefson, 1996). A persistent corpus luteum suppresses the regular sexual cycle. Western veterinarians have long used this method to induce heat in cows. To bring sows into heat, Taiwanese farmers apply acupuncture as a simple, cheap, natural and effective technique (Hsia and Lee, 1989). Conversely, stockraisers may sometimes manipulate tissue in such a way as to forestall or postpone pregnancy so that it will occur at a more convenient time or when forage and water conditions are optimal. For instance, according to Bekalo (personal communication), camel-raisers insert stones into the uterus of their female camels to prevent them from getting pregnant.

## The cultural and socio-economic context

Stockraisers do not necessarily share the same notions of microorganisms and cells as Western scientists. For example, when rumen pH is too high or too low, they do not speak of the microflora and -fauna of the intestinal tract but rather of an "imbalance" in the animal's body or a "sourness" in its belly. For these or other ailments, people may also mention supernatural etiologies, such as a punishment from the gods, evil eyes, spirits, or even black magic on the part of an enemy. Or they may cite natural, but invisible or unknown, agents such as winds, vapours or other phenomena that can contaminate pastures, water sources and animal quarters. Sometimes they say changes in the weather are to blame, and so forth. These conceptualisations (some of the more naturalistic of which are essentially accurate) do not in themselves invalidate the empirical basis or the practical effectiveness of the many beneficial veterinary practices developed by stockraisers.

Of course, the concepts and rationale behind such practices vary from society to society, as also do the particular form the practices take. To illustrate with bloodletting, stockraisers of one culture may explain that, in bleeding the neck vein of an ailing animal, they are expelling harmful spirits (a supernatural explanation). But people of a different culture may instead bleed the ear with the stated aim of removing poisons (a more naturalistic explanation). In other words, there is no single ethnoveterinary system worldwide, in the same way that one can speak of Western or scientific veterinary medicine. Rather, ethnoveterinary medicine is an integral part of a people's sociocultural and biophysical environment. As such it is greatly influenced by cultural beliefs concerning magic, religion and cosmology; by societal norms and trends; and by the materia medica available in the form of local flora, fauna and minerals, which vary according to climate and other factors.

The next sections review disease causation beliefs (ethnoetiologies) and classification systems (ethnotaxonomies) in a cultural context before addressing the following questions: Which social groups hold what kind of ethnoveterinary knowledge? How is this knowledge acquired and shared? When do people decide to treat a sick animal instead of merely selling or slaughtering it? And when or why do they choose traditional instead of modern treatment options, or something in between the two?

## Disease causation and classification

Two broad types of human or animal disease etiology can be distinguished in nearly all cultures: supernatural and natural. The former explain illness by reference to agents such as gods, genies and evil spirits, or to magical procedures. The latter see health problems as the result of a disturbed physical equilibrium. Examples of naturalistic medical systems are humoral pathology, India and Sri Lanka's Ayurvedic medicine, the Chinese medical system, and the hot/cold theories prevalent in Latin America and much of Africa.

Supernatural and natural explanations of illness are not mutually exclusive. Many cultures recognise both. Beni-Amer pastoralists in Ethiopia provide an example. They class livestock diseases into several broad categories representing a mix of supernatural and natural causes: diseases that result in sudden death are explained in terms of "destiny" and "supernatural will"; other categories, transmissible, chronic and curable/preventable diseases, are defined by hot/cold and other naturalistic features of the ailments they embrace and by stockraisers' ability to treat or control them.

Ethnoveterinary systems generally name and classify most diseases by their salient clinical signs, sometimes combining these with epidemiological observations and supernatural considera-

with epidemiological observations and supernatural considerations. As a result, ethnotaxonomies sometimes lump together scientifically different diseases, especially when the ills in question share similar syndromes and epidemiological profiles. Among the Quechua, for instance, a single term has been found to span as many as eight scientifically distinct conditions. Conversely, a disease considered as single by scientists may fall into more than one local category. For example, coastal farmers in Kenya consider East Coast Fever a distinct ailment in juvenile rather than adult cattle (Delehanty, 1996). Similarly, farmers in Africa and elsewhere may consider FMD to be two distinct diseases, depending on whether the lesions manifest themselves on the foot or the mouth.

The fact that ethnoscientific classifications differ from Western scientific ones does not necessarily mean they are wrong, nor always less discriminating. The treatments applied may work nevertheless, and ethnotaxonomies can outstrip Western ones in their complexity. To take but one example, camel pastoralists of the Western Sahara have at least 600 terms covering the diseases, reproductive physiology and anatomy of their camels (Monteil, 1952, cited in Köhler-Rollefson, 1996). It is clear, however, that translating local veterinary taxonomies and vocabulary into their Western-scientific correlates is no easy task. Ideally, it should be undertaken by a team of biological, linguistic and social scientists.

## Generation, distribution and transfer of ethnoveterinary medicine

As in the generation of other kinds of local technical knowledge, ethnoveterinary understanding, practices and skills are built up over time from empirical observation, mainly through trial and error and sometimes through deliberate, or even desperate, experimentation and innovation. For instance, Quechua who find their beloved herds of llama and alpaca threatened by a fatal disease will experiment with one treatment after another until they hit upon one that works, hopefully before the entire herd succumbs to the disease. An especially astute example of stockraiser invention is the adoption by Baggara Arabs of an early commercial tissue-culture vaccine that provided two years protection against rinderpest. They combined use of the vaccine with their own ethnoveterinary knowledge of immunisation; towards the end of the second year, they mixed their commercially inoculated cattle in with other, rinderpest-infected stock, correctly calculating that the resulting mild infection would confer permanent immunity to the disease and at no additional cost!

The successful results of such experimentation and invention often enter the broader ethnoveterinary toolkit, as stockraisers exchange information and experience among themselves. Their techniques may even come to form part of practical Western veterinary medicine, recommended by clinical veterinarians to other

cheap and efficient technique for preventing copper deficiency in ruminants; the animal is made to swallow a copper penny, which lodges in the rumen where it slowly oxidises. Vets in Britain now sometimes recommend this simple and inexpensive technique to their clients (Schillhorn van Veen, personal communication). If, as has often been noted by students of the sociology of knowledge, yesterday's science may become today's superstition, it is also true that yesterday's ethnoscience may becomes today's conventional science.

Of course, ethnoveterinary knowledge and skills are not always evenly spread within a community or a society. Their distribution across and among different social, economic, gender, caste and other groups is influenced by several factors.

First is the level of knowledge involved. Knowledge that is relatively simple and easily put into practice may be held by most stockraisers. However, the more complex and specialised it is, the fewer the individuals who may be party to it (Warren, 1992). For instance, all adults in an Andean agropastoral community may know at least one generalised treatment for diarrhoea among their herds; some will know of more treatments, and of husbandry practices that can help prevent the problem or its recurrence; but

Many stockraising groups routinely perform at-home post mortems when their animals die of disease. Here, two Andean stockraisers search for the definite diagnosis of an ailment that has attacked their flocks of sheep. Lacking access to microscopes, lab, scientific texts and often even any formal schooling, people often have recourse to practical necropsy as one of their main means of garnering empirical, ethnodiagnostic information. (*Constance McCorkle*).

only a few are able to link specific treatments to specific types of diarrhoea in different species or classes of livestock. A general rule of thumb appears to be that basic husbandry practices and remedies for very common diseases are more widely known and more often applied than are more complicated vaccines, drugs and surgical procedures. Particularly for surgery, the attention of local experts or specialists is usually required.

Looking across cultures, a second determinant in the distribution of ethnoveterinary knowledge and skills emerges. This is the degree to which people rely upon livestock for their survival. In the forests of the humid tropics, where stockraising is less important than cultivation or other productive activities, such knowledge and skills are usually scarce. Whether in the Andes or Africa, pastoralists have a much more profound and widely shared corpus of ethnoveterinary knowledge than do mixed farmers or agropastoralists. In Kenya, for example, the Intermediate Technology Development Group (ITDG) has documented the astute diagnostic abilities, the many effective remedies and the impressive surgical skills of Gabra, Pokot, Samburu and Turkana pastoralists. In contrast, among settled cultivators such as the WaKamba and Meru, only a few specialists hold most of the ethnoveterinary information, which itself is also diminished to only half a dozen or so remedies per specialist. Such findings had immediate implications for the design of ITDG paraveterinary training programmes in Kenya. With pastoralists, it proved more feasible to train a large number of herders to treat their own animals, which was the pastoralists' preference anyway; but for farming communities it was more effective to train a few, selected paravets (Grandin et al., 1991; Iles, 1994; Iles and Young, 1991).

A third distributional factor is the division of pastoral labour and related tasks by gender, age or other biosocial parameters (McCorkle, 1994b). In Andean households, for instance, it is often men who see to the daily grazing of cattle, while girls and boys, supervised by adult women, herd the family's small ruminants (sheep, goats, llama and alpaca) in the fields and rangelands, and women tend the guinea pigs, swine and poultry at home. Among the Samburu in Kenya, males usually see to the slaughter of healthy livestock, that is culling, whereas females slaughter and butcher the sick and expired animals; females are also responsible for cleaning and cooking certain internal organs from the diseased animals (Iles, 1994). In addition, Samburu women do most of the collection and preparation of plants for use in indigenous veterinary drugs (Stem, 1996). Among Gabra pastoralists, only uncircumcised boys are allowed to milk camels; young men are generally in charge of the cattle; and while anyone can milk the goats, women are the recognised experts when it comes to the delicate task of milking sheep (McCorkle, 1994a).

As a rule, those social groups that deal most frequently or intimately with a given species or with particular aspects of livestock management and utilisation have a greater or more precise corpus of ethnoveterinary knowledge in these areas. This link between the division of labour and the distribution of knowledge and skills has obvious implications for accurately recording and assessing the latter. If the relevant group is not consulted on the topics in which they are the experts, incomplete or wrong data may result.

This is what happened on the Small Ruminant Collaborative Research Support Program in the Andes, which initially addressed its animal health care enquiries and activities only to men. However, the men showed little interest in the many visible health problems of their family's small ruminants. Researchers finally realised that women were the ones responsible for these species. Only when the programme began to involve the women was the true extent of local knowledge revealed (McCorkle, 1989). An ethnoveterinary research project among the Samburu fell into a similar error. At first, the project conducted its interviews on indigenous herbal drugs only with men. Later, the research team discovered that women collected and prepared most of these medicines. When the women were subsequently interviewed, the team learned that many of the men's responses had been inaccurate or inappropriate (Stem, 1996). Yet another project discovered that Samburu women were also more knowledgeable about certain endoparasites of livestock, due to their role in slaughtering and processing diseased animals (Iles, 1994).

A fourth factor affecting the distribution of ethnoveterinary knowledge is occupational specialisation. In Nigeria, for example, hunters who use dogs are better informed about canine ethnomedicine; similarly for racehorse owners and equine health problems (Ibrahim, 1986). In Sri Lanka, elephant owners and mahouts, who make their living from hiring out and training/managing these great beasts, are more knowledgeable in elephant health care (personal communication, 1994).

Now, how do these different groups acquire, conserve and transfer their local knowledge across time and space? The answer: in some instances, through written records; but mostly through non-formal education and local/indigenous communication networks, including ceremonial and ritual events; and, of course, through application (Mathias, 1994). Apparently no study has yet been conducted focusing on the transmission of ethnoveterinary knowledge *per se*, although snippets of information can be found in a few publications. From the authors' field experiences, it appears that most stockraisers' education about animal husbandry and health care takes place mainly verbally and experientially, "on the job" in and around the farm or camp, rather than through structured and systematic teaching.

A partial exception to this rule are certain veterinary traditions with their own training centres, universities and research institutes, and with written records dating back in some instances as much as 4000 years. These include the Chinese, Ayurvedic and Greco-Arab Unani/Tibb traditions (cf. Anjaria, n.d.; Schwabe, 1996). Sri Lankan veterinary lore for elephants was written down on leaves several centuries ago and then transferred to print in the late 19th century (Piyadasa, 1994). And whether for Sri Lankan elephants or for the camelids of the ancient civilisations of the Andes, some of their latter-day owners have kept meticulous personal journals of local treatments (personal communication, 1994; McCorkle, field experiences in Peru 1980-90). Widespread interest in systematically documenting the ethnoveterinary knowledge of peoples outside such "high civilisations" as those just listed has arisen only within the last two decades.

At the local level the dominant mode of transmission of ethnoveterinary knowledge is from father or mother to child. For example, the son of a now deceased goat castrator in Java described how his father had learned his skills from his grandfather (Mathias-Mundy et al., 1992). And a renowned Samburu bonesetter-surgeon-obstetrician explained how, to pass on his art, he would select from among his sons "the one with the greatest aptitude" (McCorkle, 1994a). Similar inter-generational transfer is reported for much of Asia (personal communication, 1994; Piyadasa, 1994), parts of Europe (for example Brisebarre, 1996), and throughout Africa. In Africa, transfer is less rigid, with instances of grandparents, uncles, in-laws and even neighbours passing on their ethnoveterinary lore to apt and interested younger male and female relatives and community members (e.g. Anonymous, 1991 and 1992; ITDG, 1989).

The latter sources also note supernatural mechanisms for acquiring ethnoveterinary knowledge and skills, such as through dreams or contact with spirits. This was the norm among North American Indians, for whom a spiritual experience with an individual animal of the species to be treated was a further means of transmission. Such mechanisms have been best documented for the Amerindian "horse doctors" of the North American plains (Lawrence, 1996). Finally and curiously, the literature is silent on the possibility of ethnoveterinary education via formal apprenticeship to a recognised healer, as is common for other traditional professions such as blacksmithing, crafting and trading.

Channels for communicating ethnoveterinary information among healers and/or stockraisers are largely the same as those for local communication generally: folk media such as songs, chants, drama, dance and ritual; local organisations of many sorts; economic and other activities; and written records (McCorkle et al., 1988; Mundy and Compton, 1992).

A classic example of the use of folk media is the Horse Dance and its accompanying songs and chants, as traditionally performed by Crow and Blackfoot Amerindians in their horse-medicine societies. The Horse Dance was a ceremony for initiating new members into the society and instructing them in its secrets of equine health care, husbandry, magic and ritual; longstanding members also took this opportunity to compare notes and exchange information (Lawrence, 1996).

Javanese villagers say they often learn the details of livestock remedies from neighbours, whether through engaging in general farm talk with them, posing specific questions, or simply watching what they do (Mathias-Mundy et al., 1992). Worldwide, daily activities provide opportunities for exchanging ethnoveterinary information, for example, while attending livestock markets, waiting in line to water one's animals at a well, organising cooperative work parties for various husbandry tasks (round-ups, shearing, castrating, branding, dipping, and so on), or simply chatting outside the mosque/church, or in other gathering places.

As in other domains of indigenous/local knowledge, ethnoveterinary skills are kept alive by people practising them. Practical application contributes to the transfer of local knowledge in at least two ways. It helps people remember the details of, for example, a certain drug preparation or a surgical operation, enabling them to transmit this information more accurately and easily to others; and it provides opportunities for demonstrating a practice to others. The Samburu healer mentioned above, for instance, always welcomes all interested parties to watch him at work and to discuss or even note down details of his procedures. Indeed, in the present and many other researchers' experience, people are often eager and proud to discuss or demonstrate their ethnoveterinary savvy. This attitude has been documented among many ethnic groups, including Javanese and Laotian farmers, Quechua agropastoralists, at least eight distinct pastoral groups in Africa, and several African cultivator groups. But not all individuals or societies are so free with their ethnoveterinary knowledge. We have already mentioned the Plains Amerindians' secretive horse medicine. In present-day Kenyan farming communities too, traditional healers of livestock are wary of sharing their knowledge with co-villagers and outsiders (Anonymous, 1991), as also are Sri Lankan specialists in elephant medicine, who pass their art on from father to son (Piyadasa, 1994).

When, why and with whom a secretive attitude toward ethnoveterinary knowledge prevails seems to depend on several factors. Almost everywhere, ordinary stockraisers appear to engage in farm talk and the exchange of health care experience and opinion among themselves and with outsiders. The situation with recognised healers, however, can vary. One hypothesis to explain this is as follows: where ethnoveterinary knowledge is substantial and widespread (as among pastoralists) or where there are few specialised healers (as among Quechua agropastoralists), information circulates more freely. But where there is an established cult or a tradition of specialists who earn a significant income from treating livestock, information is guarded more jealously. As with many other aspects of ethnoveterinary knowledge and practices, however, more research on this question is needed.

## Decision-making in ethnoveterinary medicine

How do people decide whether to sell, slaughter or treat a sick animal and, in the last case, whether to employ traditional or modern medicine? The literature is largely silent on these questions. Perhaps this is because stockraisers' decision-making in this regard is a balancing act among many factors. Also, it is arguably more complex than taking decisions on human patients, where slaughter is not an option, although neglect may be.

At a minimum, the factors involved include the type and progression of the disease, and the relative expense in both cash and opportunity costs of accessing ethno- versus conventional veterinary services, as against simply slaughtering the sick animal. With regard to the latter option, consumption and marketing considerations will also figure in decision-making; for example: household methods for preserving the meat of large stock, the distance to market and the transport costs entailed, local demand for meat, government price policies and livestock taxes, meat safety regulations, and slaughterhouse fees. Cross-cutting all the foregoing factors are: the general economic value of the animal(s) in question (for example a goat is usually worth less than a cow); and the animal's specific economic value (that is whether it has passed its peak of growth, milk production, fertility, or traction and transport service). People's assessment of specific value will also vary according to their priority production goals for the species in question (dairy, meat or traction, or absolute numbers of animals), as well as the shifting market prices of different products. Any special status attaching to the patient may have to be considered too. For example, the ailing individual may be the most reliable lead animal for the herd, the best-trained or strongest work animal, the only suitable stud to which the household has access, and so on. Still other considerations may enter into the sell, kill, or cure decision, such as an owner's psychological or personal magical involvement with the patient, or religious taboos.

In most stockraising societies known to the authors, people tend to have favourite animals. A Dutch colleague described how his mother had a favourite cow for whom she always saved the choicest kitchenscraps and whose trip to the slaughterhouse was postponed well beyond its economically optimal milk life (Schillhorn van Veen, personal communication). In both the Andes and Africa, an individual's luck, health and well-being in life may be linked in magico-religious ways to the fate of an animal assigned to him/her at birth. As an example of a taboo, devout Hindus will under no circumstance slaughter sick or for that matter, healthy cattle, although purposive neglect is a "treatment" option. Among Muslim stockraisers in Nigeria, in part because of religious concern over slaughter methods, economic returns have been shown to vary greatly according to the point in disease progression at which an ailing animal is sold for meat. If the creature is still in apparent good health, it can be sold to an itinerant merchant for 100% of its meat-and-hide value. If it is visibly ill but still able to walk as far as the local butcher, it will fetch on average 75% of this price. But with or without treatment, if the animal becomes too sick to move and must be slaughtered at home by undocumented methods, the meat will earn only 25% of its value (Schillhorn van Veen, personal communication).

A further example of how ideology or cosmology can figure in veterinary decision-making is provided by Quechua agropastoralists. Even when hard-nosed economics would seem to dictate slaughter, they will spare no expense to try to save one of their precious camelids. Llama and alpaca are surrounded by a wealth of myth and semi-sacred proscriptions and prescriptions as to their·care and handling. They are said to have been given to humans by the great Earth Mother who, if she sees they are being mistreated, will take them back again. Without camelids, humans could never survive in the high Andes.

In certain parts of Ireland, farmers seldom consult a veterinarian for sheep, but they are quick to do so for cattle. Although there appears to be no specific ideological reason for this difference, it cannot be explained solely by the greater economic value of cattle. During the 1930s, government policies in Ireland caused cattle to fall in value until they were almost worthless, yet the demand for bovine veterinary services did not slacken (Shanklin, 1996). Health care decisions for the two species appeared to relate more to their role in the whole farming system than to market prices.

A similar case is reported among smallholders in Brazil's semi-arid Sertão. Here, little labour, capital, veterinary or other care is devoted to the region's skinny but hardy goats; such inputs are instead reserved for the household's more lucrative but ecologically risky cattle, sheep and cropping enterprises. The role of goats in Sertão farming systems is to act as a last-ditch, low- or no-cost emergency backstop to cover basic family needs for cash and food when other animals and crops succumb to one of the region's severe and recurrent droughts. Making veterinary or other investments in goats would defeat people's purpose in keeping them in the first place (McCorkle, 1996; Primov, 1989).

Javanese villagers also make very pragmatic veterinary decisions about their small ruminants. They try treating an ailing animal themselves before taking any other measures. Their decision to sell or slaughter is influenced by whether the disease is chronic or acute and directly life-threatening. For a slowly progressing skin disease like scabies, most villagers prefer to sell rather than slaughter an animal if their home remedies fail; but for an acute and often fatal condition such as poisoning, their decision is the other way around (Wahyuni et al., 1992).

For cases in which stockraisers do decide to treat their animals, data on when and why they choose traditional versus modern veterinary medicine, assuming both are available, are exceedingly rare. One study on Ethiopian cattle raisers found that, out of 101 respondents, 25 used traditional treatments, 11 modern, and 6 both, the remaining 59 did nothing. However, these choices did not correlate either with the type of disease or with household income. This study also found that Ethiopian farmers chose traditional medicine more frequently for livestock than for humans. This finding was attributed to three factors. First, the farmers were all fairly knowledgeable about ethnoveterinary treatments and readily shared their knowledge, whereas this was not so for human ethnomedicine. Second, they had less access to formal veterinary assistance than to modern medicine for humans. Third, people were more willing to pay for expensive commercial drugs for ailing family members than for livestock (Ghirotti, 1996).

To summarise, it is almost impossible to draw cross-societal generalisations concerning stockraisers' decisions to treat, sell or slaughter. However, further research might confirm various broad hypotheses that are hinted at in a critical reading of the literature. One is that, barring psychological and ideological considerations, where stockraisers know the disease in question to be almost invariably fatal no matter what veterinary treatments are used, they logically choose slaughter. Another is that, once a decision to treat is taken, a rough rule of thumb may be: try home remedies first; if these don't work, try the next cheapest option, such as a local livestock healer or a paraveterinarian; finally, seek fully professional veterinary attention if the animal is of great economic or production-system value, either as a last resort or for diseases that one believes only Western veterinary medicine can cure.

## Commercialisation of ethnoveterinary medicine

In parts of Asia, stockraisers have one further treatment choice: to purchase and themselves administer manufactured versions of ethnoveterinary medicines. Most examples of this option come from India, where between 60 and 80 small firms presently produce trademarked veterinary drugs based on traditional botanicals (Anjaria, 1996). One or two Indonesian pharmaceutical houses also sell traditional medicines for livestock.

There are both advantages and disadvantages to commercialising traditional drugs. Among the latter, first is the thorny issue of intellectual property rights. Veterinarians have been known to collect prescriptions for home remedies from stockraisers and then sell or report them to drug companies. Second, as noted earlier, traditional polyprescriptions are difficult to validate scientifically. Third, maintaining quality control in production can be a problem. Fourth is the question of purchase price. In India, manufacturers' profit margins on herbal drugs are high – smallholders would probably be ahead financially if they prepared crude plant materials themselves at home rather than buying ready-made drugs.

The latter point is moot, however. The present authors know of no studies on the relative opportunity costs of stockraisers purchasing such manufactured drugs or preparing their own prescriptions at home. Assuming both types of remedy are equally effective, their relative economic advantages would depend on considerations such as the work-time and cash outlays involved in travel to gathering sites versus marketplaces, plus the time needed to prepare and administer the two types of drug, compared with the local opportunity cost of labour.

What is certain is that, in cash terms, proprietary herbal medicines produced in-country are generally cheaper than the equivalent Western drugs. Other advantages in establishing a domestic industry in traditional medicines are that it can provide local or regional income and employment opportunities, and it can help keep alive valuable local knowledge that might otherwise be lost to encroaching Westernisation.

# Importance and future of ethnoveterinary medicine

## Limitations and potentials

In the light of the foregoing, what can be said about the limitations and potentials of this rural people's science? And what gaps in research need to be filled before its full potentials can be realised?

Ethnoveterinary medicine has a number of limitations. Obtaining, preparing or administering some home remedies may be inconvenient and time-consuming, although stockraisers may voice the same complaints about Western pharmaceuticals. Many medicinal plants are seasonally unavailable, making treatment impossible if they must be used fresh. Effective ethnotherapies for infectious epidemic diseases are for the most part lacking, although prophylaxis may be possible. Also some traditional treatments are ineffective or downright harmful. Moreover, stockraisers' diagnoses are often inadequate, resulting in ineffective therapy or therapy that comes too late to save the animal. Lastly, standardising local vaccines and drugs and determining optimal dosages and treatment schedules can be difficult.

Despite these caveats, ethnoveterinary medicine has great potential. As indicated earlier, many of its techniques are familiar to Western-style veterinarians. As for unfamiliar practices, one must beware of dismissing them out of hand merely because they are couched in a non-scientific idiom. Folk practices that sound bizarre to Westerners may in fact have practical medical value and constitute rational adaptations to local infrastructural and economic conditions. (Recall the example of the copper penny or, to take another example: to increase the fertility of female animals, following the law of signatures that like affects like, stockraisers may feed certain plant parts because these resemble the shape of the female sexual organ.) Given the relationship between fertility and adequate nutrition, such dietary supplements can nevertheless promote fertility. Similar practices have been demonstrated as effective in human ethnomedicine (Etkin, 1990). Yet another example is the belief, held by many herders, that places where highly contagious diseases have broken out are inhabited by evil spirits and should therefore be avoided. For diseases with long-lasting pathogens (like *Clostridia*), this is just another way of encoding valuable epidemiological information and one that leads to practical and effective management action.

Of course, as noted earlier, some traditional treatments can be ineffective or even harmful. But the same can be said of Western veterinary treatments if they are applied ineptly or when contraindicated. As we have seen, many other ethnoveterinary techniques in fact make sound scientific sense.

Effective ethnoveterinary remedies and practices offer several advantages over corresponding Western-style drugs: they are generally cheaper, more accessible and more readily understood, sometimes more environmentally friendly, often socioculturally more acceptable and better adapted to local realities (McCorkle, 1995).

The use of botanicals and other traditional medicines seems to be especially appropriate for wound care, parasites and skin diseases. Indigenous husbandry practices such as feeding regimes and basic surgical and obstetrical practices also appear broadly applicable. Traditional vaccinations are probably best limited to poxes, FMD, and ecthyma; where available, commercial vaccinations are generally more appropriate for other diseases.

A common misperception is that reliance on ethnoveterinary medicine is restricted mainly to remote rural peoples. At least one study has shown that peri-urban stockraisers use it too. Respondents from three villages close to the Javanese city of Bogor reported that local agents of the government veterinary service hardly ever visited them. Moreover, respondents said that the time, distance and travel costs of reaching a professional veterinarian precluded them from accessing such services (Mathias-Mundy et al., 1992). The actual extent and use of ethnoveterinary medicine differ according to a variety of factors, but it is mainly practised by individual stockraisers and traditional healers at the community level or, in the case of renowned healers, at a regional level. Outside India, the commercial-industrial production of herbal medicines for animals is still limited.

The potential benefits of ethnoveterinary medicine are outlined in Box 3.2. Fully realising these potentials will require further research and development, organised technology transfer, and forward-looking policy reform.

## Future directions

### Research

Recommendations for research on ethnoveterinary medicine involve both laboratory and field studies.

---

**Box 3.2**
**Potentials of ethnoveterinary medicine (EVM)**

- Recognising, validating, developing and transferring EVM gives stockraisers more technology choices
- With more and cheaper veterinary options to choose from, greater improvements in animal health can be realised. This in turn may produce:
- Increased income (in cash or kind) due to reduced need to purchase animal products
- Increased income from other sources that rely on healthy work animals for traction or transport
- Increased food security, due to the foregoing and also due to increased numbers or improved condition and hence value of livestock
- Improved human nutrition
- Increased status, prestige and stored social capital from raising more and healthier animals

**Box 3.2
continued**

- Decreased risks to the environment and to livestock and human health, incurred by avoiding the use of toxic commercial products*
- Increased local or national interest in preserving the native flora and fauna used in materia medica and special breeds of local livestock, and hence the maintenance of biodiversity
- Renewed respect among local/indigenous people for their own culture and technological expertise
- Empowerment of people to take greater control of their lives by virtue of their increased or more confident knowledge base
- Increased respect on the part of practising veterinarians, development workers and scientists for their clients' and beneficiaries' rich fund of local knowledge and experience.

* For example, by substituting highly toxic and biocumulative dips, which may poison people and water supplies, with tightly targeted topical applications and/or husbandry measures.

Source: McCorkle and Mundy.

The efficacy of ethnoveterinary drugs needs to be scientifically validated and documented. At present, only a handful of drugs have been properly studied. The results of the few studies conducted are often published as theses, working papers and other "grey" documents, and are therefore difficult to access. As a starting point, it would be useful to collect and summarise such studies in order to make their findings more widely known. Next, descriptions of ethnoveterinary medicine should be screened for remedies that sound effective. Selected remedies should then be tested in the laboratory and, more important, on the animal species in question.

Several plants used in ethnoveterinary medicine have already been studied in the context of human ethnomedicine. In such cases, data on an ingredient's efficacy *in vitro* will be valid for veterinary medicine, with the result that laboratory tests need not be repeated. Efficacy *in vivo* does differ, however, rendering additional testing on animals necessary. Different species may react differently to the same remedy and will require different dosages, treatment schedules and application methods.

While testing, researchers should bear in mind that a remedy's efficacy is determined not only by a plant's active ingredients but also by its collection and preparation methods and the combination of plant species and parts used. A useful starting point is to follow local methods as closely as possible, determining dosages and treatment schedules for different animal species and classes. Research in the field should aim first to prioritise the diseases and problems for which to test remedies. A field study in Java, for example, indicated that scabies was a major problem in goats, while footrot rarely occurred under traditional management (Wahyuni et al., 1992). Thus testing remedies for scabies should have priority over footrot and other diseases.

We mentioned above that dosages and treatment schedules are often weak points in ethnoveterinary systems. However, studying local approaches to disease control may provide important clues for the design of veterinary interventions and projects. Pastoralists in West Africa, for instance, are reluctant to dose their whole herd regularly with anthelmintics, as Western veterinary medicine recommends. Rather, they treat only animals that are clinically ill, typically during the rainy season when parasite burdens intensify. A double-blind study measuring treatment costs and cattle performance in semi-arid Niger demonstrated that this approach makes good sense both economically and in terms of parasite challenge. Clearly, Nigerian herders are more likely to accept strategic therapeutic measures than ponderous and costly prophylactic schemes, so veterinary research and development efforts targeting such groups should be designed accordingly.

Stockraisers' veterinary decision-making is another area requiring research. How do stockraisers decide whether to sell, kill or cure a sick animal? When do they use traditional and when modern veterinary medicine? Answers to these questions can be crucial for the design of veterinary interventions. The data currently available do not allow us to do more than hazard a few broad hypotheses, as outlined above.

## Technology transfer

Despite their considerable potentials, ethnoveterinary practices have found little place in technology transfer, particularly in government extension services. This can be attributed partly to the lack of guidance on what practices are appropriate. Without such guidance, development professionals hesitate to integrate ethnoveterinary practices into project design and implementation or into extension recommendations.

To provide some direction on the use of ethnoveterinary medicine, in July 1994 the International Institute of Rural Reconstruction (IIRR), in collaboration with Heifer Project International, held a workshop to compile an ethnoveterinary information kit for South-East Asia. Participants included Asian farmer-stockraisers and Asian and other scientists. The kit specifies remedies that all in the group agreed they could recommend to animal health care workers and stockraisers generally. It serves to increase recognition of the value of ethnoveterinary medicine and to make it more accessible. Perhaps more importantly, this pioneering effort offers a model for the development of other country- or location-specific kits for direct use at the farm level.

## Development

During the past few years, the international development scene has undergone significant changes. Governments, donors and mainstream development organisations have come to realise that local participation is essential for successful projects and programmes with lasting benefits. Field research and the planning and implementation of development activities are increasingly drawing on methods such as participatory rapid appraisals (PRA) and participatory action research (PAR).

Such methods have been successfully used in the study and application of ethnoveterinary medicine. For examples, see IIED (1994) and McCorkle and Bazalar (1996). However, much remains to be done before real and widespread stockraiser participation at the field level can be said to have been achieved. The term, participatory, has become fashionable and many field studies and projects have been generously labelled with it. Ideally, local people should be the main actors, while outsiders confine themselves to a supportive role. Yet a careful look at, so-called, participatory activities often reveals that, in fact, local people's participation consists merely of answering questions, with little or no say in decision-making. Such shortcomings are often inadvertent, due to a lack of knowledge and experience. Guidelines, case studies and especially training could help government extensionists and the field personnel of non-government organisations (NGOs) become more adept at participatory project planning and implementation. Such tools are particularly useful in development efforts that aim to work with and build on local knowledge, of which ethnoveterinary medicine is an important branch.

## Policy

Many international and national organisations have yet to recognise the potential contribution of ethnoveterinary medicine to development. This stands in marked contrast to human ethnomedicine, which has been widely acknowledged and used by development organisations. Experience shows that official recognition and support for traditional medicine can create a climate favourable to its promotion. The Philippine Ministry of Health, for instance, has endorsed the use of 10 traditional plants for human treatments. This has facilitated the development of community-based processing centres for remedies made from these plants.

However, both NGOs and scientists interested in the production of plant medicines have noted the need to address quality standardisation and product pricing, among other issues. Existing government regulations are not generally suitable for small-scale community-based enterprises that produce herbal remedies. On these and other issues, more policy work is needed to create an enabling environment for the practice and dissemination of ethnoveterinary medicine.

## Outlook

Integrating ethnoveterinary information and practices into research and extension services can make national livestock development efforts more successful. Ethnoveterinary medicine can be useful whenever and wherever stockraisers have no other animal health care options, whether in rural or in peri-urban areas. It can even help where conventional veterinary services and commercial drugs are available, by providing cheaper alternatives to expensive Western drugs for selected diseases, thereby expanding the range of treatments available to users.

The key to success is to find the right blend of Western and local practices and remedies. Unfortunately, there is no single recipe for doing this. Guiding considerations should include whether local remedies are effective and easily available; how much labour is necessary for their preparation and administration compared with the costs of commercial drugs; and whether the remedies are environmentally friendly and safe. A prerequisite for success is that all involved – stockraisers, development professionals, academics and politicians – should acknowledge the value of ethnoveterinary medicine as a real alternative or complement to Western medicine. Only then will we able to select and combine the best of both worlds to the advantage of rural people.

## Notes

(1) Due to constraints on the length of this chapter, it does not reference the several hundred documents cited in McCorkle (1986), Mathias-Mundy and McCorkle (1989), or McCorkle and Mathias-Mundy (1992). The interested reader is encouraged to consult these sources directly for complete references to the information and examples given here, as well as for many further examples of the techniques described in this chapter. In contrast, references from the forthcoming McCorkle et al. (1996) volume on ethnoveterinary research and development are cited individually, so as to acknowledge the contributions of the many dedicated scientists and development workers who participated in that effort.

## References

Anjaria, J. n.d. 'Legendary of ethnoveterinary medicine in some Asian countries', unpublished ms

Anjaria, J. 1996. 'Ethnoveterinary pharmacology in India: Past, present, and future' in C.M. McCorkle, E. Mathias and T.W. Schillhorn van Veen (eds), *Ethnoveterinary Research and Development,* Intermediate Technology Publications, London

Anonymous. 1991, 'Traditional Akamba veterinary skills,' progress report of research carried out at Kibauni Location of Mwala Division in Machakos District, January 1991, unpublished manuscript, Intermediate Technology Development Group, Nairobi, pp. 137–147

Anonymous. 1992, 'Report on discussions with traditional healers, Mtito Andei, October 1992,' unpublished manuscript, Intermediate Technology Development Group, Nairobi

Baekbo, P. and Nielsen, K. 1992, 'Immunization of pregnant sows by feeding faeces from piglets with diarrhea: Attempt to prevent diarrhea in suckling pigs,' paper presented at the 1992 International Pig Veterinary Society, cited in *Updates* (Jan): 8–9

Berge, E. and Westhues, M. 1969, *Tierärztliche Operationslehre,* Paul Parey, Berlin

Bodeker, G.C. 1994, 'Traditional health care and public policy: Recent trends,' *Nature and Resources* 30 (2): 5–16

Brisebarre, A-M. 1996 'Tradition and modernity: French shepherds' use of medicinal bouquets,' in: C.M. McCorkle, E. Mathias and T.W. Schillhorn van Veen (eds), *Ethnoveterinary Research and Development,* Intermediate Technology Publications, London

CUSRI/CESO. 1988, 'Reports on group discussions, in Indigenous Knowledge and Learning,' papers presented at the Workshop on Indigenous Knowledge and Skills and the Ways They are Acquired, Cha'am, Thailand, 2–5 March 1988. Chulalongkorn University Social Research Institute, Bangkok, Thailand, and Centre for the Study of Education in Developing Countries, The Hague, Netherlands, pp. 123–133

Delehanty, J. 1996, Methods and results from a study of local knowledge of cattle diseases in coastal Kenya,' in C.M. McCorkle, E. Mathias and T.W. Schillhorn van Veen (eds), *Ethnoveterinary Research and Development,* Intermediate Technology Publications, London, pp. 229–245

deMaar, T.W. 1992, 'Ask what's in those bottles,' *Ceres:* 24 (4): 40–45, Food and Agriculture Organisation, Rome

Etkin, N. 1990, 'Ethnopharmacology: Biological and behavioral perspectives in the study of indigenous medicines,' in T.M. Johnson and C.F. Sargent (eds), *Medical Anthropology: Theory and Method*, Praeger, pp. 149–158, New York

Farnsworth, N.R. 1983, 'The NAPRALERT data base as an information source for application of traditional medicine,' in Robert H. Banneman, J. Burton and C. Wen-Chieh (eds), *Traditional Medicine and Health Coverage: A Reader for Health Administrators and Practitioners*, Geneva," World Health Organization, pp. 184–193

Foster, G.M. and Anderson, B. 1978. *Medical Anthropology*, Alfred A. Knopf, New York

Ghirotti, M. 1996, 'Recourse to traditional versus modern medicine for cattle and people in Sidama, Ethiopia,' in C.M. McCorkle, E. Mathias and T.W. Schillhorn van Veen (eds), *Ethnoveterinary Research and Development*, Intermediate Technology Publications, London, pp. 15–53

Grandin, B., Thampy R. and Young, J. 1991. *Village Animal Healthcare: A Community-based Approach to Livestock Development in Kenya*, Intermediate Technology Publications, London

Hsia, L.C. and Lee, J.H. 1989, 'Inducing oestrus by acupuncture,' *Animal Husbandry and Agriculture Journal* (April): 18–21

Ibrahim, M. A. 1986, 'Veterinary traditional practices in Nigeria,' in R. von Kaufmann, S. Chater and R. Blench (eds), *Livestock Systems Research in Nigeria's Subhumid Zone*, proceedings of the Second ILCA/NAPRI Symposium held in Kaduna, Nigeria, 29 October–2 November 1984, International Livestock Centre for Africa, Addis Ababa, Ethiopia

Ibrahim, M.A. and Abdu, P.A.. 1996, 'Ethno-agroveterinary perspectives on poultry production in rural Nigeria,' in C.M. McCorkle, E. Mathias and T.W. Schillhorn van Veen (eds), *Ethnoveterinary Research and Development*, Intermediate Technology Publications, London

IIED. 1994, RRA Notes No. 20, Special Issue on Livestock, International Institute for Environment and Development (IIED), London

IIRR. 1994, *Ethnoveterinary Medicine in Asia: An Information Kit on Traditional Livestock Healthcare Practices*, Vols 1–4. International Institute of Rural Reconstruction, Silang, Cavite, Philippines

Iles, K. 1994, 'The role of ethnoveterinary work in the ITDG/OXFAM Animal Health Programme, Baragoi, with reference to pastoralists' perceptions of helminths,' unpublished consultancy report, Intermediate Technology Development Group, Nairobi

Iles, K. and Young, J. 1991, 'Decentralised animal health care in pastoral areas,' *Appropriate Technology* 18 (1): 20–22

ITDG. 1989, 'Traditional healers workshop of the Intermediate Technology Development Group Utooni Development Project,' 30–31 March 1989, unpublished manuscript, Intermediate Technology Development Group, Nairobi

Köhler-Rollefson, I. 1996, 'Traditional management of camel health and disease in North Africa and India,' in C.M. McCorkle, E. Mathias and T.W. Schillhorn van Veen (eds), *Ethnoveterinary Research and Development*, Intermediate Technology Publications, London, pp. 129–136

Lawrence, E.A. 1996, 'I stand for my horse: Equine husbandry and healthcare among some North American Indians,' in C.M. McCorkle, E. Mathias and T.W. Schillhorn van Veen (eds), *Ethnoveterinary Research and Development*, Intermediate Technology Publications, London, pp. 60–75

Mares, R.G. 1954, 'Animal husbandry, animal industry and animal diseases in the Somaliland Protectorate,' Part II, *British Veterinary Journal* 110 (11): 470–481

Marx, W. 1984, 'Traditionelle tierärztliche Heilmethoden unter Berücksichtigung der Kauterization in Somalia,' Giessener Beiträge zur

Entwicklungsforschung: Beiträge der klinischen Veterinärmedizin zur Verbesserung der tierischen Erzeugung in den Tropen, Reihe I, Band 10: 111–116, Wissenschaftliches Zentrum Tropeninstitut, Justus-Liebig Universität, Giessen, Germany

Ma'sum, K. 1991, 'Traditional veterinary medicine for ruminants in East Java', in E. Mathias-Mundy and T.B. Murdiati (eds), *Traditional Veterinary Medicine for Small Ruminants in Java.* Bogor, Indonesian Small Ruminant Network, pp. 29–37, Bogor, Indonesia

Mathias, E. 1994, 'Indigenous Knowledge and Sustainable Development,' Working Paper No. 53, International Institute of Rural Reconstruction, Silang, Philippines

Mathias-Mundy, E. and McCorkle, C.M. 1989, 'Ethnoveterinary Medicine: An Annotated Bibliography,' *Bibliographies in Technology and Social Change* No. 6. Iowa State University, Ames, Iowa

Mathias-Mundy, E. and Murdiati, T.B. (eds) 1991, *Traditional Veterinary Medicine for Small Ruminants in Java*, Indonesian Small Ruminant Network, Bogor, Indonesia

Mathias-Mundy, E., Wahyuni, S., Murdiati, T.B., Suparyanto, A., Priyanto, D., Isbandi, Beriajaya and Roemantyo, H.S. 1992, 'Traditional animal health care for goats and sheep in West Java: A comparison of three villages,' SR-CRSP/Indonesia Working Paper No. 139. Small Ruminant Collaborative Research Support Program, Balai Penelitian Ternak, Bogor, Indonesia

Matzigkeit, U. 1990. *Natural Veterinary Medicine:Ectoparasites in the Tropics,* Verlag Josef Margraf, Wikersheim, Germany

McCorkle, C.M. 1986, 'An introduction to ethnoveterinary research and development,' *Journal of Ethnobiology,* 6 (1): 129–149

McCorkle, C.M. 1989, 'Toward a knowledge of local knowledge and its importance for agricultural RD&E,' *Agriculture and Human Values* 6 (3): 4–12

McCorkle, C.M. 1995, 'Back to the future: Lessons from ethnoveterinary RD&E for studying and applying local knowledge,' *Agriculture and Human Values* 12 (2): 52–80

McCorkle, C. M.1996. 'The roles of animals in cultural, social and agro-economic systems,' in V. James (ed), *Sustainable Development in Third World Countries: Applied Theoretical Perspectives,* Greenwood Press, Westport, Connecticut

McCorkle, C.M. 1994a, 'Ethnoveterinary R&D and Gender in the ITDG/Kenya RAPP (Rural Agricultural and Pastoral Development Programme),' unpublished consultancy report, Intermediate Technology Development Group, Nairobi

McCorkle, C.M. 1994b, 'A Framework for Analysis of Gender and Other Socio-economic Variables in Ag&NRM,' Working Papers on Women and International Development No. 241, Women in Development Program, Center for International Programs, Michigan State University, East Lansing, Michigan

McCorkle, C.M. and Bazalar, H. 1996, 'Field trials in ethnoveterinary R&D: Lessons from the Andes,' in C.M. McCorkle, E. Mathias and T.W. Schillhorn van Veen (eds), *Ethnoveterinary Research and Development,* Intermediate Technology Publications, London

McCorkle, C.M., Brandstetter, R.H. and McClure, G.D. 1988, *A Case Study on Farmer Innovations and Communication in Niger,* Academy for Educational Development for AID/S&T and USAID/Niger, Washington, DC

McCorkle, C.M. and Mathias-Mundy, E. 1992, 'Ethnoveterinary medicine in Africa,' *Africa* 62 (1): 59–93

McDonald, P., Edwards, R.A. and Greenhalgh, J.F.D. 1973. *Animal Nutrition,* 2nd. ed. The English Language Book Society and Longman, London

Mundy, P. and Compton, L. 1992 'Indigenous communication and indigenous knowledge: Concepts and interfaces,' paper presented at the International Symposium on Indigenous Knowledge and Sustainable Development, held at the International Institute of Rural Reconstruction, Silang, Cavite, Philippines, September 20–25, 1992

Piyadasa, H.D.W. 1994, 'Traditional systems for preventing and treating animal diseases in Sri Lanka,' *Revue Scientifique et Technique de l'Office Internationale des Epizooties* 13 (2): 47–486

Plotkin, M.J. 1988, 'Conservation, ethnobotany and the search for new jungle medicines: Pharmacognosy comes of age...again,' *Pharmacotherapy* 8: 257–262

Primov, G. 1989, 'The role of goats in agropastoral production systems of the Brazilian Serto,' in C.M. McCorkle (ed), *Plants, Animals and People: Agropastoral Systems Research,* Westview Press, pp. 51–58, Boulder, Colorado

Roepke, D.A. 1996, 'Traditional and re-applied veterinary medicine in East Africa,' in C.M. McCorkle, E. Mathias and T.W. Schillhorn van Veen (eds), *Ethnoveterinary Research and Development,* Intermediate Technology Publications, London, pp. 256–264

Schillhorn van Veen, T. 1996, 'Sense or nonsense? Traditional methods of animal disease prevention and control in the African Savanna,' in C.M. McCorkle, E. Mathias and T. Schillhorn van Veen (eds), *Ethnoveterinary Research and Development,* Intermediate Technology Publications, London, pp. 25–36

Schwabe, C. 1996, 'Ancient and modern veterinary beliefs, practices and practitioners among Nile Valley peoples,' in C.M. McCorkle, E. Mathias and T.W. Schillhorn van Veen (eds), *Ethnoveterinary Research and Development,* Intermediate Technology Publications, London, pp. 37–45

Shanklin, E. 1996, 'Care of cattle versus sheep in Ireland: Southwest Donegal in the early 1970s,' in C.M. McCorkle, E. Mathias and T.W. Schillhorn van Veen (eds), *Ethnoveterinary Research and Development,* Intermediate Technology Publications, London, pp. 179–192

Stem, C. 1996, 'Ethnoveterinary R&D in production systems,' in C.M. McCorkle, E. Mathias and T.W. Schillhorn van Veen (eds), *Ethnoveterinary Research and Development,* Intermediate Technology Publications, London, pp. 193–206

Trotter, R.T. 1988, Seminar on Mexican Ethnomedicine, 20 October 1988, Iowa State University, Ames, Iowa

Wahyuni, S., Murdiati, T.B., Beriajaya, Sangat-Roemantyo, H., Suparyanto, A., Priyanto, D., Isbandi, and Mathias-Mundy, E. 1992, 'The Sociology of Animal Health: Traditional Veterinary Knowledge in Cinangka, West Java, Indonesia: A Case Study,' SR-CRSP/Indonesia Working Paper No. 127, Small Ruminant Collaborative Research Support Program, Balai Penelitian Ternak, Bogor, Indonesia

Warren, D.M. 1992, 'Training materials on indigenous knowledge,' International Institute of Rural Reconstruction, Silang, Philippines

WHO. 1991, 'Report of the Consultation to Review the Draft Guidelines for the Assessment of Herbal Medicines,' World Health Organization, Munich, Germany

# 4 Biopesticides

*Saleem Ahmed and Gaby Stoll*

## Introduction

An estimated one-third of global agricultural production, valued at several billion dollars, is destroyed annually by over 20 000 species of field and storage pests (McEwen, 1978; Agrios, 1978). The synthetic pesticides widely used for control are valued for their effectiveness (although in many cases this diminishes over time), their relatively long shelf life (when properly stored), and the ease with which they can be transported, stored and applied. However, they also cause serious problems. Foremost among these are toxity, with an estimated three million or more cases of pesticide poisoning annually, 20 000 of which prove fatal (Dinham, 1993); pollution of soils, water and air, with as yet unknown long-term consequences for humans, wildlife and the environment; and the development of pesticide resistance, necessitating the use of larger and larger doses at increasing cost to the farmer and to society as a whole. In South Asia today, cotton must be sprayed 15 to 16 times a season, compared with five or six times 10 years ago.

Alternative, low-cost pest and disease control strategies are thus urgently needed. This need is doubly felt by resource-poor farmers in the marginal production areas of developing countries, who have not generally benefitted from the Green Revolution. Plant materials with appropriate properties may play an important role in the development of such strategies, especially materials which can be harvested, formulated and used by farmers, applying their own knowledge and skills. In fact, until recently many such materials were widely used in traditional societies in developing countries, although they remain little known to Western science. The plant species that are the source of such materials may thus be thought of as the "unsung heroes" of traditional approaches to pest and disease control. The purpose of this chapter is to explore and promote their possible future use.

## Basic concepts

### Non-plant materials and strategies

Plants are not, of course, the only source of biopesticides. Nature provides a wealth of other renewable resources which have been found useful for pest and disease control. In fact, the choice of materials and strategies is probably limited only by human ingenuity and willingness to experiment, plus the time and resources with which to do so.

As regards materials, several non-botanical but locally available substances are used for pest and disease control in some countries. For example, a slurry of cow's urine and manure is commonly applied to the floor of village huts and granaries in India. Farmers on the island of Leyte, in the Philippines, burn scraps of old car tyres in their fields.

A major strategy used in developing countries and also gaining ground in the developed world is the package of practices commonly referred to as organic farming. This usually includes composting, crop diversification, crop rotation, following a crop calendar, and hand weeding. Using these practices, organic farmers in Japan are reportedly obtaining yields of paddy and vegetables similar to those obtained by "conventional" farmers, who use very high amounts of chemical fertilisers and synthetic pesticides (Ahmed, 1994).

Another strategy, now the subject of intense research, is that of biological control. This consists of using natural enemies, setting one organism against another (Mackauer et al., 1990). Perhaps the best-known example is the successful use of the parasitoid wasp *Epidinocarsis lopezi* to control the cassava mealybug in tropical Africa (IITA, 1990). Other species of parasitoid wasp have also been successfully used to control the brown planthopper (*Nilaparvata lugens*) in paddy rice and various stemborers in sugarcane. Similarly, the microbial pesticide *Bacillus thuringiensis* is currently being tested in several countries to control the diamondback moth (*Plutella xylostella*), which is a serious pest of vegetables, and the maize stemborer. Pathogenic organisms have also been used to control weeds, notably *Lantana camara* in Hawaii.

Other strategies recommended as components of what is known as integrated pest management (IPM) include the use of resistant crop varieties, "clean" cultivation, baits and traps, and the planting of decoy crops.

Many of these non-plant materials and strategies are used in combination with preparations made from plants. In traditional societies plants were and in some cases still are the main source of biopesticides. It is on these plant sources that the rest of this chapter will focus.

## The plant as biopesticide

Why doesn't the rice stemborer lay its eggs in the cotton boll? And why doesn't the cotton bollworm lay its eggs in the papaya fruit? The questions asked by today's scientists and farmers, in their pursuit of sustainable pest control strategies, are not essentially different in kind to those posed by early agriculuralists, although they may be couched in different terms.

The association of pests with specific plant hosts is usually governed by a physical attribute of the plant and/or by its biochemical composition. Thus, if we could simulate cotton's environment in the rice field, for example by spraying an appropriate cotton extract onto the rice crop, we could conceivably keep out the rice stemborer. Theoretically, any plant species not attacked by a specific pest could provide a biopesticide to control that pest on other crops. Since this does not have to be a killing action, it could just as well be a repellent, antifeedant, or growth regulatory action, we could replace the term biopesticide with botanical pest control (BPC) agent. The potential for discovering much needed BPC agents is vast, since modern science has taken no more than the first few steps in exploring this field.

## Traditional uses

It is difficult to draw the line between traditional and modern uses of BPC agents, since several so-called modern applications are merely traditional ones that have been scaled up and developed commercially. A later section will deal in more detail with botanicals that have been subject to recent (post World War II) research and development.

### A rich and renewable resource

According to Grainge and Ahmed (1988), more than 2400 plant species around the world are currently known to possess pest control properties. While the listing of these authors was based on reports in the scientific literature, most of the studies that gave rise to these reports were inspired by local knowledge and practices. Many such studies were conducted by developing country scientists, who are well placed to tap local and often anecdotal information found in the folk lore or in traditional rural areas. Indeed, research on botanical materials for pest control has continued in the developing countries, whereas in the industrialised North it declined with the advent of synthetic pesticides. Growing concern over the harmful effects of chemical control is currently triggering renewed interest in BPC agents worldwide, but since the pesticide industry generally develops only materials suitable for large-scale commercial synthesis, most of the plants on the list remain chronically underresearched.

The BPC species identified belong to 235 plant families differing greatly in their plant cycle, habitat and growth characteristics, as well as in the pests they control or are alleged to control, the type of pest-control activity they exhibit, and their complementary uses. They range from tiny mosses and lichens to giant trees; from grasses and sedges to legumes; from desert cacti and succulents to tropical rain forest trees and shrubs; and from plants found in the frigid polar regions to those native to the torrid equatorial zones

(Appendices Table 1). The family Asteraceae has, thus far, been found to contain the largest number of BPC species, with 261.

In addition to this global review, several national studies on traditional uses have been carried out. In Thailand, for example, Chejew et al. (1988) has documented alternative pest management knowledge and practices throughout the country, producing a list of commonly used plants, together with the plant parts used, the mode of action and the target pests (Table 4.1).

TABLE 4.1   Plants commonly used as biopesticides in Thailand

| Plants | Parts used | Major mode of action | Target pests |
|---|---|---|---|
| Acorus calamus | rhizome | contact poison repellant | ants |
| Allium sativum | pulp | fungicide | Pseudoperorosora cubensis |
| Alpinia galanga | rhizome | repellant contact poison | Dacus dorsalis Nilaparvata lugens, Aphid rot, brown spot |
| Annona muricata | leaf | contact poison | Phyllotrita sinuata, P. punchata |
| A. squamosa | leaf, seed | contact poison | body louse, P. sinuata |
| Azadirachta indica | leaf, seed | anti-feedant growth regulator repellant ovicidal | sucking insects, butterflies, aphids larvae |
| Curcuma domestica | rhizome | contact poison repellant fungicide | red mite, thrip ants, Xanthomonas citri Anternaria porri |
| Cypobogon nardus | leaf | repellant | Dacos dorsalis, mosquito |
| Derris elliptica | stalk, root | contact poison | various insects |
| Dioscorea hispida | tuber | contact poison | various insects |
| Euphorbia fringona | whole plant | contact poison | various insects |
| Nicotiana tabacum | whole plant | contact poison | larvae aphids |
| Stemona tuberosa | leaf, root | contact poison | various insects |
| Tinospora crispa | vine | anti-feedant | diamond-backmoth |
| Toona tomentosa | | contact poison growth regulator | N. lugens, Aphid sp., butterflies |
| Zingiber zerumbet rhizome | | contact poison | aphids, larvae butterflies |

Source: Chejew et al. (1988).

## A science develops

Knowledge of the medicinal properties of plants probably pre-dates agriculture. It is thought to have begun when human beings were still hunters and gatherers, inhabiting the primary forests and their associated savannas.

Trees and forests were, and in some cases still are, central to traditional societies as their primary source of food, shelter, medicine and other products and services. They have also been vener-ated and worshipped. In India, for example, Aranyani, the Goddess of Forests, and Vana Durga, the Goddess of Trees, are still worshipped as the primary sources of life and fertility. In the Andes and Africa too, nature is venerated as sacred. Within the sacred forest, special areas are dedicated to local deities, and trees within these enclosures are regarded with great respect. If a tree was cut down, another is planted in its place. The diversity, harmony and self-sustaining nature of the forest are thus the organisational principles guiding society, which take the forest as its evolutionary model. Human evolution is measured in terms of people's capacity to merge with nature's rhythms and patterns, intellectually, emotionally and spiritually. The centuries-old lifestyle of such societies should not, therefore, be regarded as primitive but rather one of conscious choice (Shiva, 1988).

Being derived from the living forest, indigenous forestry science did not perceive trees merely as wood. Instead they were seen as multi-purpose, with a focus on diversity of form and function. It was in the context of this worldview that the systematic knowl-edge of plants and plant ecosystems that characterises traditional societies gradually arose. As it accumulated, this knowledge was transmitted from generation to generation. In some societies it became so abundant that it was vested in a specialised person, the medicine man or woman. Then, when agriculture began, a further specialisation developed, the art of using certain plants or plant parts to protect food crops from pests and diseases.

*Pyrethrum, a biopesticide, is a crop threatened by biotechnology research.*   Harvest of Nature's Diversity, Ceres 151: p. 6.

Scientists first learned of the pest-control properties of plants by observing the practices of traditional farming societies. Many examples of pest control using plants are found in ancient litera-ture. The earliest known mention of poisonous plants occurs in the Rig Veda, the classic book of Hinduism, composed in India during the second millennium B.C. (Chopra et al., 1949). The ancient Romans used false or white hellebore (*Veratrum album* or *V. viride*) as a rodenticide, and the Chinese are credited with dis-covering the insecticidal properties of *Derris* species (Feinstein, 1952). In A.D. 970, the Arab scholar Abu Mansur listed 584 natural materials possessing pharmacological and poisoning properties (Achundow, 1889/1968). Pyrethrum was used as an insecticide in Persia and Dalmatia in the eighteenth and nineteenth centuries (Nelson, 1975), and tobacco plant preparations have been similar-ly used in the Middle East for nearly two centuries (Frear, 1943). According to Secoy and Smith (1983), approximately 700 plant species have reportedly been used over the past few decades in different parts of the world for pest control, regardless of efficacy. Plant-derived chemicals, such as pyrethrum, rotenone and nicotine, were used for pest control in the West for decades, but lost out to synthetic pesticides after World War II.

## A regional overview

Despite the growth of the modern agro-chemicals industry, many plant species are still used for pest control in the developing world. Traditional applications are found in every developing region, but they are better documented in some regions than in others.

In Africa, for example, several Nigerian tribes are known to mix the dried leaves of a wide range of plants with stored grain to control post-harvest pests. The species used in this way include *Annona senegalensis* Pers., *Luffa aegyptiaca* Mill., *Hyptis spicigera* Lam., *Ocimum americanum* L., *Afromosia laxiflora* Harms., *Erythrophleum guineense* G. Don., *Butyrospermum parkii* Kots., *Datura stramonium* and *Nicotiana* spp. Stems of *Ocimum americanum* are similarly used, while in the case of *Lantana rugosa* Thimb. the whole plant is employed (Giles, 1964). As in other regions, many of the trees used in Africa are multi-purpose; in Zimbabwe, sap from the bark of *Spirostachys africana* serves as a pesticide in granaries, while the wood is sculpted or used as lintels and the tree as a whole is thought to ward off evil spirits.

The Central and South American Indians have long used the mamey tree (*Mammea americana*) for pest control purposes (Pagon and Morris, 1953), and even today it is a common practice in Puerto Rico and elsewhere in the region to use whole mamey leaves as wrappings around newly set plants to prevent insect attack at or below ground level (Duarte and Franciosi, 1976). In other cases, ground mamey parts are applied as dust, or mixed with water or kerosene and sprayed. Other commonly used biopesticides in Latin America are *Allium sativum, Azadirachata indica, Capsicum frutescens, Gliricidia sepium, Luinus mutabilis, Melia azedarach* and *Nicotiana tabacum*.

In West Asia to North Africa, the nicotine-rich water from hookahs (water pipes) is often applied to plants as a crop protection measure. The use of botanical remedies against crop pests and diseases would seem to be less common in this region, perhaps because constraints of this kind are less severe owing to the dry conditions.

The use of plants to control pests is widespread and well documented in South and East Asia. On the island of Leyte, in the Philippines, rice farmers insert twigs of *madre de cacao* (*Gliricidia sepium*) between every four hills of paddy to control stemborer, while other farmers spray a paste made from hot peppers (many times diluted) to control foliar pests on a range of crops (Ahmed et al., 1988). Pastes are also made from seaweed or tree resin. In Nepal's Kathmandu valley, farmers used to poke twigs of *Artemisia* spp. into paddy fields to prevent stemborer attack, a practice which was subsequently abandoned in favour of chemical pesticides because these chemicals were more effective and available at inexpensive (subsidised) prices (Ahmed, 1984). To control pests in stored grain, people in South Asia either simply mix a handful of neem leaves with the grain, or apply a neem leaf paste as a coating on the inside of the earthen container used for storage (Ahmed and Grainge, 1986). Dried neem leaves are also placed in books, carpets and clothing to protect against silverfish, termites and other pests. For nematode and termite control in cardamom, fruits and vegetables, farmers in south India incorporate 100-250 kg/ha of neem cake or *Pongamia glabra* cake annually into the soil, while to control crop pests they spray aqueous neem leaf, *Pongamia glabra* leaf extract or diluted neem oil (Ahmed and Grainge, 1986). In Timor, grain stored in baskets woven of *Corypha utan* palm leaves reportedly suffers less rodent damage than when stored without this form of protection (C. Lamoureux, personal communication).

Lastly, China deserves a special place in the annals of botanical pest control. Among the 267 plant species used in this country are *Tripterygium wilfordii,* aqueous extracts of which are sprayed to control caterpillars, tussock and pine moths, and mustard, melon and rice leaf beetles; *Stellera chamaejasme,* whose roots are ground into a powder which is then mixed with topsoil to control soil pests; and *Melia azedarach,* dried leaf blades of which are placed between the mat and supporting board of beds to control lice and fleas (Yang and Tang, 1988).

## Processing and trade

Except in the case of neem, little information is available on efforts at village level to harvest, process and market plant materials for pest control. Farmers often collect these materials free of charge from trees growing in and around their village and use the plant parts directly, without any processing.

In many cases, however, it is primarily the women's task to gather plant parts and prepare home-made remedies. It is safe to say that in most developing countries women are both the prime repositories and the prime users of knowledge on these matters. Frequently, they are assisted by the elderly and by children. Collecting and preparing botanical pest control materials are light activities that can, for the most part, be performed at any convenient time, thus taking their place among many other tasks. Harvesting activities include plucking leaves, flowers or fruits from the plant, or sweeping the dropped fruit from the ground (as in the case of neem and *Pongamia glabra*). Preparation activities include cleaning, winnowing, drying, crushing, grinding or soaking in water. Women may also be responsible for application, or for selling small surpluses on to traders or processors.

In the case of neem cake, a sophisticated collection, processing and distribution system is in place in several countries of South Asia. People sweep neem fruit from under the trees, wash, depulp and dry it, and then sell it either to neem seed traders or directly to neem crushing facilities, thereby supplementing the family income. Village traders, in turn, sell it either to crushing units, to wholesalers, or to a few very large neem soap manufacturers. Crushing units with capacities of between five and twenty tonnes of neem seed per day are now widespread in rural India. They are operated by an assortment of private or public non-government organisations (NGOs) and cooperatives. Their main product is neem seed oil, which is used, either by the same units or by others, to produce neem soap. This commands a premium on the market because of its purported germicidal properties. Neem cake, produced as a by-product, is used as a soil amendment and for fertiliser coating. The growing demand for neem cake for nematode and termite control in south India may be gauged from the fact that 3000 tonnes of this product are currently sold annually in the region's "Cardamom Hills", much of it by pesticide dealers (Ahmed and Grainge, 1986).

## Levels of use

Most of the surveys so far conducted on traditional plant-based BPC agents have been qualitative rather than quantitative, trying to understand what plants are used and how, rather than the percentage of farmers using them and the effectiveness of different applications. As a result, hard data on levels of use, as on other important factors, are generally lacking.

The exception to this state of affairs is neem, for which several authors have conducted more detailed studies. Ahmed (1988) reports that approximately 50% of the 500 or so farmers surveyed in India and Pakistan who store foodgrains for more than six months use neem leaves to ward off storage pests. Users are generally the poorer and illiterate farmers, who obtain the leaves free of charge from trees in their backyard or in the neighbourhood. In contrast, the use of neem cake for nematode control is more common among relatively affluent and better educated farmers; 17 out of 19 cardomom growers surveyed in south India applied 100–250 kg neem cake/ha to control nematodes. It appears that a fairly large percentage of south India's vegetable growers also use this product. Farmers buy neem cake in the market because they find no synthetic pesticide to be equally effective.

It is difficult to extrapolate from these figures to the population as a whole, since even within the same village the number of users may change over time. However, it would seem safe to assume that neem leaf users in these countries as a whole must run into the millions.

Similar information needs to be collected for other botanical pest-control materials. While the use of neem is widespread in several countries, that of most other species appears ad hoc and highly localised. As a result, extrapolation will be even more problematic.

# Recent research and development

## Progress since World War II

The rise of the agro-chemicals industry since 1945 led to the near total demise of traditional BPC agents in the industrialised North. In the South, synthetic pesticides were increasingly used during the 1950s and 60s, at first by large-scale farmers on cash crops only, but then by small-scale farmers also, on a range of food and cash crops. However, because these products remained expensive or unavailable in many areas, their use in the South has been less pervasive than in the North. Extensive use and overuse are still restricted to certain countries and regions, generally the higher-potential areas of Latin America and Asia.

By the end of the 1970s, agriculture in these areas was plagued by rising levels of resistance in insect pests, the pollution of soil and water resources and the declining health of farmers and their families. First cotton, then a number of other crops, had suffered yield crashes as pests became resistant to all known chemical products, with devastating economic consequences for farmers. Researchers and NGOs, particularly the latter, began looking for alternatives to synthetic pesticides.

In the past 15 years, NGOs in several developing countries have conducted surveys on indigenous pest control methods. The countries covered include Mexico (Rodriguez et al., 1984), Guatemala (Munos, 1985), the Philippines (ATCRD, 1984), Thailand (Chejew et al., 1988), Senegal (Thiam and Ducommun, 1993) and Zimbabwe (Elwell and Maas, 1995). To these surveys should be added the literature reviews and surveys initiated by actors in the North, including Giles (1964), Jacobson (1975), Secoy and Smith (1983), Ahmed (1984), Ahmed et al. (1988), Librero et al. (1988), Escano et al. (1988), Stoll (1992) and Berger (1994).

In the early 1980s the donor community began funding research on biopesticides at national and international research centres and universities. NGOs were little involved in this research, but a number of them went ahead with their own efforts to promote biopesticides by conducting action research with farmers, by collecting and disseminating information and by encouraging farmer-to-farmer exchanges. In some cases, such as that of neem in Nicaragua, this pioneering work led to renewed interest on the part of government and so to policy changes.

Since the 1992 United Nations Conference on Environment and Development, which emphasised the conservation of biodiversity as a major priority, donor support to the development of biopesticides has, unfortunately, slowed down.

## Impact

The impact of recent research and development efforts is impossible to guage at global level, since for many countries the necessary information is simply not available. Of necessity, then, we must follow an anecdotal approach, visiting a few countries only in each of the world's developing regions.

In 1985, no one in Latin America seemed interested in the use of BPC agents. Ten years later, natural crop protection is in fashion, apparently featuring in almost every agricultural training event. Much of the increased interest can be attributed to a demand pull from farmers, who have faced escalating costs for synthetic pesticides. However, farmers have so far learnt more from each other, through farmer-to-farmer exchanges, than from formal research and extension.

In Nicaragua today, natural crop protection is synonymous with neem. Although the plant was introduced less than 20 years ago, most small-scale farmers now have five to ten neem trees growing on their land. This is a success story for the country's NGO community and government, but the emphasis on this tree has led to the neglect of other locally available biopesticides, such as *Melia azedarach*, which were once more widely used.

Among the world's developing regions, South and South-East Asia have seen perhaps the most extensive research and development efforts in the past 20 years. Again, however, most of the effort has gone into neem, to the possible exclusion of other worthwhile candidates for research.

In Thailand, the availability of neem-based biopesticides was a prerequisite to the emergence in the early 1990s of a new market for organic vegetables, fruit and rice, now widely sold through a network of "green" shops. The first neem-based product was developed by a farmer, after which an NGO, Technologies for Rural Ecological Enrichment (TREE), adapted it for use in other areas through participatory action research (see Box 4.1). By 1990 a commercial company had been set up to develop and market biopesticides. The company has demonstrated that a national biopesticides industry can be viable, providing a useful model for other countries to follow (see Box 4.2). The Thai experience has influenced government policy in neighbouring Laos, which is now also taking steps to promote neem.

The Vietnamese Ministry of Agriculture launched a programme on a range of alternative plant protection methods in 1989. The programme conducted a survey on farmers' use of plants with insecticidal properties. This led to the establishment of three new botanical gardens containing 53 plant species with a potential for greater use. Among these, priority is given to the legume *Derris elliptica*, widely grown by farmers in the south of the country and a well known botanical insecticide in other parts of the world (see Box 4.3).

**Box 4.1**
**Thailand: NGO builds on farmers' inventiveness**

In 1985 Khun Annop, a farmer-entrepreneur, developed his own neem-based biopesticide. He wanted to protect his five hectare orange grove from pests and diseases. This was the start of what has now become a flourishing small-scale national industry.

TREE began developing neem-based biopesticides with farmers in 1989, as part of a national project on alternative crop protection. The project operated near Suphanburi, in the "rice bowl" of Thailand, an area where Green Revolution technology had caused rising levels of resistance in insect pests, including brown planthopper. Mr Annop's spray had contained *Citronella* grass, which did not grow in this area. Farmers were encouraged to develop and apply their own criteria for identifying possible alternatives. Using a traditional classification of plants

as hot, bitter, poisonous or smelly, they were able to short-list a number of species for experimentation.

The research led to several new insights and adaptations of the technology for specific circumstances. During an outbreak of brown planthopper, one farmer noted that neem-based products were ineffective where rice plants were dense. This led to a switch to a derris-based product that acted as a contact insecticide. Another farmer experimented by broadcasting halved citrus fruits at the same time as spraying with a neem-based product. Ingredients successfully used elsewhere, such as tobaco leaves or rhizomes of *Curcuma domestica*, were tested for their efficacy against specific pests. The labour requirement was reduced by making bigger batches for storage in barrels, and by eliminating the chopping of raw materials.

Several researchable issues were brought to the attention of

the formal research system. Perhaps the most significant was the effectiveness of neem as a fertiliser as well as an insecticide. This was based on farmers' observations that rice was greener where neem-based products had been applied. Despite fruitful cooperation with the local Farming Systems Research Institute, it proved difficult to persuade research groups to take up research on biopesticides. One solution would be to encourage NGOs to contract research to the formal system.

To disseminate the technology, the project enlisted the help of 20 core farmers to act as resource persons. By 1993, they had introduced over 1000 other farmers to neem-based pesticides, several hundreds of whom are now thought to be regular users. Further technology transfer is taking place through the informal networks on biopesticides that have developed among local NGOs and farmers.

Source: Srithong et al. (1994).

---

The commercial potential of traditional BPC agents appears relatively low in Africa, where farmers still have less cash to spend than in Asia. In 1993 the francophone NGO Environonnement Africain (ENDA) held a workshop to explore West African experiences with natural crop protection. The resulting proceedings (Thiam and Ducommun, 1993) describe a range of traditional uses and modern research efforts, concluding that several applications show sufficient efficacy to warrant production on a larger scale. One example is a paper describing an account of Senegalese farmers' experiments comparing a synthetic pesticide, Cyperm thrine, with biopesticides made from *Allium sativum* and *Melia azederach* on a range of crops including tomato. The authors conclude that the biopesticides are generally as effective as the synthetic pesticide. *Allium sativa* even proved more efficient at controlling grasshoppers (*Kraussaria angulifera*). Other plant species investigated by scientists in the region include *Hyptis spicigera*, *Cassia nigricans*, *Ocimum basilicum*, *O. gratissimum*, *Cymbopogon schoenanthus* and *Boscia senegalensis*, besides the opportunities for using neem. The lists of plants

TREE staffmember Phayong Shrithong with a billboard which can be placed in rice fields as a tool to inform farmers. It reads: Botanical Pesticides: true friend of the farmers. Control and eliminate Brown Plant hoppers, diamond black moths, stemborers and other insects resistant to pesticides with the silent but effective power of biopesticides. Farmers can produce it themselves. Safe to the user, non-destructive to the environment. Supported by TREE. (*Wim Hiemstra*).

and the ways of preparing them were long, but the active substances, mode of action and efficacy remained unknown in most cases, indicating a tremendous knowledge gap for research to fill.

## Farmers' constraints

Many farmers are aware of the pest-control properties of some plant species, but do not use them. Why not? Listed below (not in any particular order) are some important reasons for non-use revealed by farmers in the course of surveys:

● The stigma of "backward practice". Neem and *Pongamia glabra* leaves for pest control in stored grain are, as we have seen, widely used by resource-poor farmers in India and Pakistan. However, their more affluent neighbours generally prefer synthetic pesticides. While some non-users were unsure of neem's efficacy, others feared being thought "backward" for using such a traditional practice. Farmers using modern inputs are considered "model farmers" by government research and extension services. It is on their farms that Agriculture Department personnel lay out crop demonstration plots and hold field days, to which neighbouring farmers and visiting

Fermentation of neem-based biopesticides proves to be more beneficial, for Thai rice farmers, than only controlling insects. It also has fertilising properties and the fermentation may also positively influence soil life.
(*Wim Hiemstra*).

An orchard farmer, near Suphanburi, Thailand, explains how he started experimenting with the neem-based biopesticide in a small area.
(*Wim Hiemstra*).

**Box 4.2**
**Natural Plants Limited**

Established in 1989, Natural Plants is a registered company committed to sustainable rural development in Thailand. Its first product was Neemix, a neem-based pesticide sold as a powder. Subsequent research led to the development of a still more effective concentrate solution, marketed as Neemix Super. This product has proved popular with commercial vegetable, flower and fruit producers and currently accounts for 90% of the company's turnover. It is 50% more expensive than Neemix, which continues to be used by urban gardeners and small-scale farmers. New by-products pioneered by the company and launched in 1994 included a neem oil shampoo for dogs and an emulsion to control external parasites.

The secret of the company's success lies in its investment in research and development and its insistence on quality control, which have led to its survival while others have fallen by the wayside. In the early 1990s neem became a popular substitute for chemical pesticides. The market was flooded by unscrupulous operators selling inferior products that had not been properly tested. Farmers soon became disillusioned, and the market collapsed. Now only two or three companies are still in business.

Neem's poor reputation among farmers has also hurt sales of Neemix, but the company is seeking to overcome its problems through direct marketing, in addition to retailing through "green" shops and conventional pesticide stores. Thailand's flourishing market in organic produce looks set to give the company a bright future. The immediate aim is to double the production capacity, which will help to lower prices and so reach a larger market, especially resource-poor farmers.

Source: Lianchamroon (1994).

dignitaries are invited. Similarly, farmers surveyed on Leyte, in the Philippines, mentioned that they use *Gliricida sepium* after sunset on their crops, so as to avoid ridicule by their neighbours (Ahmed et al., 1988).

● Lack of official recommendation. Many farmers follow recommendations made by the government extension service. The use of botanical materials for pest control is, however, not among these recommendations. For example, one of only two non-users of neem cake for nematode control among 19 farmers/organisations surveyed in south India was the Tea Research Institute of the United Planters' Association of South India (UPASI). Its manager indicated that they do not use the product as it is not on the official list of recommended inputs. Cases of this kind are common in many countries, as scientists continue to devote much of their time and energy to the investigation of synthetic pesticides.

● Unavailability of materials. In some cases, botanical pest control materials may no longer be available. For example, farmers near Hyderabad, India and Faridpur, Bangladesh said that all neem trees in the area had been chopped down. Scarcity and high price of neem cake were the reasons cited by other non-users in south India. Other materials are available only seasonally. Some users considered this a major constraint, while others were apparently able to plan ahead, building stocks to last through the cropping season.

● Lack of immediate "knock-down" and "total kill". Synthetic pesticide users often conclude that botanical materials do not work, as they continue to see the target pest in their fields after

**Box 4.3**
**Vietnam: Public-sector research reveals a neglected potential**

In Vietnam's Mekong Delta, derris *(Derris elliptica)* is traditionally grown on sandy soils close to the sea. Often, it is intercropped with vegetables or cereal crops, serving to reduce the labour requirement for weeding during the first few months of the season. Used as a fish poison as well as an insecticide, derris is a lucrative cash crop, fetching US$ 0.54–0.82 per kg of root material, which contains 3-6% of the active ingredient rotenone.

Under the project, farmers in nine provinces were encouraged to grow derris. The emphasis was on areas in the north of the country, where the crop was unknown. Within three years, an estimated 144 ha were being cultivated. Cultivation techniques, seed multiplication and optimum harvesting dates were studied.

To prepare for manufacture on a larger scale, the project set up two pilot production units to test different extraction methods. The best results were obtained with a wettable root powder to which small amounts of other ground plant substances had been added. Since 1989, the units have produced nearly 50 tonnes of the powder and distributed it to farmers, free-of-charge or at subsidised prices.

In field trials, applications of 3 kg/ha of the powder showed reasonable efficacy (60-70% mortality) against a wide range of pests, including diamondback moth, aphids and green leafhoppers. Applications as low as 0.7 kg/ha were reported to be as effective against the rice leaf folder as the standard synthetic pesticide in use. A hectare of derris yields about 1.5 tonnes of roots with a five per cent rotenone content – enough active ingredient to treat 500 to 700 ha of cropland.

The Ministry plans to set up more plants in the near future, this time operating commercially.

Source: Shrimpf and Pag (1993).

applying them. Even if they are aware of some effect, they feel disappointed at the lack of complete eradication.

- Frequency of use. The rapid biodegradability of botanical pesticides often means more frequent applications and hence higher labour costs. Solid or dried materials such as neem cake or *Pongamia glabra* cake, which are incorporated annually into the soil, are liked by users as they need be applied only once a year.
- High labour requirement or costs. In some cases, the time required to collect and process botanical materials can be too high for resource-poor farmers. Farm families must carefully weigh the opportunity costs of their labour before deciding whether to perform these tasks themselves. In other cases, the cost of purchasing a ready-made agent can be unacceptably high, though it is normally lower than that of synthetic products.
- Legal constraints. Some farmers surveyed in Hawaii would have liked to use crude neem materials for pest control, but were hamstrung by a "Catch 22". The regulations of America's Environmental Protection Agency (EPA) are designed to protect humans from poisonous or fraudulent materials. According to these regulations, it is unlawful to use as a pesticide any material which has not been registered with the EPA. However, natural plant extracts cannot be registered because their composition is variable, and not all the active ingredients may be known. Thus, American farmers cannot use neem extracts on produce that will be sold despite their known safety for people and the environment. Many developing countries currently base their regulations governing pesticide use on the EPA model.

● Inadequate research and extension. Following the success of the Green Revolution, most scientists devoted their time and energy to the pursuit of chemical solutions to pest problems, an approach generally endorsed, if not dictated, by policy makers. Not until the mid-1980s did the pendulum start to swing the other way, with increased emphasis on IPM approaches. With research on the effectiveness of traditional control methods out of fashion, extensionists did little to promote traditional BPC agents.

## Promoting future use

### The ideal agent

Bearing in mind the constraints facing resource-poor farmers, the plant species containing the BPC agents should ideally possess the following characteristics:

● be easy to grow, requiring little space, labour, water and fertiliser,
● recover quickly each time the pest control material is harvested from it,
● be perennial, eliminating the need for periodic replanting,
● not become a weed or a host to plant pathogens or insects,
● possess complementary economic uses,
● effectively control either a specific pest or a broad range of pests,
● pose no hazard to non-target organisms, wildlife, humans or the environment,
● be easy to harvest, formulate and use with simple village-level technology.

### Promising species

The neem tree (*Azadirachta indica*) appears to be the most promising species of all, as it possesses nearly all the characteristics mentioned above (Ahmed and Grainge, 1986). It is a hardy tree that flourishes in marginal tropical or subtropical areas with little care and few inputs. More than 15 complex chemicals (triterpenoids) having repellent, antifeedant, insect growth regulatory and pesticidal properties have been identified in aqueous and chemical extracts of its leaves, bark, stem, seed and other parts. These have been found to control more than 250 species of insects, mites and nematodes, as well as some fungal, bacterial and viral diseases. Yet they have also been found to be safe for humans and friendly to the environment. Neem leaves and other plant parts are eaten by people in India and Pakistan for numerous medicinal and therapeutic purposes.

Few other plant species possess quite as many of these desirable characteristics. Listed in Appendices Table 2 are some 50 species which come close. All of these are effective in controlling important crop pests. In nearly all cases, scientific studies to evaluate their pest control potential were stimulated by prior traditional knowledge. Many of these species are also listed in Abu Mansur's compilation dating back more than a thousand years (Achundow, 1889/1968). The species on our list grow under a wide range of climatic conditions and possess diverse types of pest-control action; methods of obtaining the pest-control formulations also vary greatly. These factors are also indicated in the table, together with the plants' complementary uses. Information on the latter and on other socio-economic aspects is still rather sketchy, as most of the work undertaken thus far has been on technical (primarily entomological and chemical) characteristics.

Distilled from Table 2 and summarised below are the salient attributes of 12 species selected for their complementary uses. We used this selection criterion because most small-scale farmers in developing countries prefer multi-purpose crops. Listed in the Appendices are the pests that each controls.

- *Acorus calamus* (sweet flag), Araceae. This is a hardy aromatic perennial herb found in a variety of tropical/subtropical, Mediterranean and temperate environments. *Acorus* is both an ornamental plant and an animal feed. Its rhizome is the source of calamus, which has medicinal and other uses besides pest control (extracts of *Acorus* are used to make perfumes). Saponins and tannins are the plant's main active principles. These exhibit pesticidal, antifeedant, repellent and attractant actions on a variety of pests; antifertility action has also been reported on rodents. Pest-control formulations are obtained by drying and powdering the rhizome and making an aqueous extract; solvents such as petroleum ether, ethyl ether and kerosene are also used in this preparation. *Acorus* extracts have been found effective in controlling ants, clothes moths, fleas, flies, fowl lice, mites, mosquitoes, moths, rats and pests of stored grain.
- *Allium cepa* (onion): Amaryllidaceae. Originating in the temperate region, this perennial herbaceous plant is now also widespread under a variety of tropical, subtropical and semi-arid conditions. It is found mainly as a cultivated plant in gardens and vegetable plots and is, as most readers know, primarily used for human consumption; but it is also used in medicinal preparations. Sulphur and tannins are its main active principles. Aqueous extracts of the leaves and bulb have been found effective against fungal diseases, mites and ticks; antifeedant and repellent actions against some insect pests have also been reported.
- *Allium sativum* (garlic): Amaryllidaceae. This is a perennial, cosmopolitan herbaceous plant found cultivated in gardens and vegetable plots. Again, many readers will be familiar with this plant, which is widely used as a cooking ingredient in human

food. The plant is also used in medicinal preparations. Its active principles are alkaloids, saponins and tannins found in the bulb and leaves. Powdering/pressing followed by aqueous, ethanol and methanol extractions are employed. The resulting formulations have antibacterial, antifungal, antinematode and antitick actions; antifeedant and repellent actions against mosquitoes, thrips and other pests have also been reported.

- *Annona reticulata* (custard apple): Annonaceae. This is a tropical or subtropical tree, grown primarily for its fruit for human consumption and for its animal feed value. It is also useful in erosion control on steeply sloping land. Although the fruit is edible, its seed contains compounds that are poisonous for humans. Hence care is needed in handling. The plant's active principles are alkaloids such as linocline and anonaine. Pest-control formulations are prepared by drying, crushing and powdering the seeds and making aqueous and alcoholic extractions. These have been found effective in controlling rice field insects and other pests.

- *Chrysanthemum cinerariifolium* (pyrethrum): Asteraceae. This is a hardy, perennial, aromatic, herbaceous plant found under diverse climatic conditions. Prized for its pesticidal properties, it also has ornamental value; however, some of its components may also be toxic to humans, domestic animals and wild life. Its dried flowers are the primary source of the insecticide pyrethrum. The alkaloid stachydrine, the plant's main active principle, is obtained by drying and powdering the flowers, followed by extraction with water, ether, alcohol, acetone or kerosene. A wide array of insects; aphids, cockroaches, flies, grasshoppers, mosquitoes, thrips and wireworms, can be controlled in this way.

- *Derris elliptica* (derris): Fabaceae. This is a perennial tropical leguminous shrub or subshrub found in Southeast Asia and other parts of the tropics. Its root is the primary source of the insecticide rotenone. The plant also has medicinal value. It should be used with caution, since some of its active principles may be poisonous to humans. Pest-control formulations, obtained by aqueous or ether extraction of the powdered root, have been found effective against a variety of insects and nematodes.

- *Lantana camara* (common lantana): Verbenaceae. This is a hardy perennial tropical hairy, sometimes prickly, shrub or subshrub growing to a height of over a metre. It is used as an ornamental cover crop, for erosion control, and as a sand binder and windbreak. It also has medicinal and feed value. However, it can grow out of hand and become a serious weed. Active principles include alkaloids (such as lantanine), flavanoids and triterpenoids. These are obtained by drying, followed by acetone and methanol extraction of stems, leaves and flowers. Several important pests can be controlled, including aphids.

- *Mammea americana* (mamey tree): Clusiaceae. This is a handsome tropical tree growing to a height of 18 metres. The tree is found mostly in the West Indies, Central America and Florida. Its fruit is used in making sweets and tropical drinks. It also has medicinal value. Its crushed seed is used as a fish poison, but may also be toxic to poultry, other animals and humans. Active principles, which are found in practically all plant parts, have pesticidal, repellent and antifeedant actions and are particularly valued for controlling ticks and mites. Formulations are made by extractions with water, alcohol, acetone and petroleum ether. A wide variety of insects, fleas, lice, mites and ticks are controlled by these formulations.

- *Ocimum sanctum* (holy basil): Lamiaceae. This is an annual herbaceous tropical plant having fragrant foliage and sweet herbs. It is used in human food, especially for flavouring. It also has medicinal value. It is held sacred by some Hindus. Its active principles are alkaloids found in roots, leaves, stems, shoots and buds. These plant parts are dried and extracted with water, ethanol or acetone. A variety of fleas, flies, maggots, mosquitoes and nematodes can be controlled.

- *Piper nigrum* (black pepper): Piperaceae. This is a perennial tropical woody climber widely cultivated in the Old World, especially in Sri Lanka, India and Indonesia, for its fruit, which is the source of the spice used for flavouring human foods worldwide. It also has medicinal value. Its main active principles are the alkaloids methylpyrroline, piperovatine, chavicine, pipieridine and and piperine. Pests controlled include fungi, insects (including mosquitoes), mites and nematodes.

- *Vitex negundo* (Indian privet): Verbenaceae. This tropical shrub or small tree is common in South-East Africa, Madagascar, India, South-East Asia, and the Philippines. It is useful as a cover crop for erosion control, and also has medicinal properties. About 12 alkaloids have so far been identified as its active principles. Formulations are obtained by crushing leaves, stems and branches and extracting with water or petroleum ether. Some important flies, clothes moths, rice insect pests, and pests of stored grain are controlled by these formulations.

- *Zingiber officinale* (common ginger): Zingiberaceae. This is a tropical herbaceous plant with tuberous and aromatic rhizomes used as the well known spice. The plant also has medicinal value. It is often cultivated in gardens and vegetable patches, and is sometimes also used as a windbreak. Aqueous extracts of its rhizomes are sprayed to control a number of important pests, and are especially valued for their effectiveness against fungal diseases.

# Research issues

## Technology

The studies so far conducted to determine the effectiveness of specific plants have been subject to several shortcomings:

- they have largely been conducted under controlled laboratory conditions,
- scientists have generally utilised chemical solvents such as acetone and alcohol to obtain plant extracts,
- they have generally looked for "killing action" in the test material, rather than the degree of plant protection obtained.

Since conditions in the field are not controlled, and since farmers do not have access to the solvents that scientists use, the first two shortcomings run the risk of over-estimating a plant's real effectiveness in the field. Conversely, because many more modes of pest control exist beside killing action (such as repellency, antifeedant effect, growth regulatory effect, etc), the third shortcoming may lead to an under-estimate of the level of pest control obtainable. We recognise that basic scientific studies need to be conducted in the laboratory. However, these studies should, when appropriate, emulate the preparation methods used by farmers, and they should always be complemented by field trials.

Other limitations concern the focus and organisation of research. The plant on which most research has been conducted is neem, which has been the subject of no less than five international conferences. Several international institutes and many national ones have projects to study its efficacy. But apart from this plant, no others have yet gained widespread attention from the research community.

The little formal research undertaken so far has been fragmented, with no joint effort behind it. No international programmes specifically devoted to plant-based biopesticides have yet been launched. The testing of BPC agents could be included in IPM approaches to pest control, but unfortunately this connection is seldom made at present. For instance, the regional programme of the Food and Agricultural Organisation (FAO) on IPM in South-East Asia has no research or training component on BPC agents. Recently, the Consultative Group on International Agricultural Research (CGIAR) has announced a global initiative on IPM, led by the International Institute of Tropical Agriculture (IITA). This represents a good opportunity to integrate research on plant-based biopesticides with other IPM components, using a networking approach. Ideally, networking would allow the use of standardised methods, as well as the exchange of materials and results.

Government extension services have so far done little to promote BPC agents. Most of their staff still consider them irrelevant and know little or nothing about them. The NGO movement has had to plug the gap, but its action research approach tends to be poorly documented and lacking in rigour, with the result that it is difficult to draw lessons useful for transferring technology elsewhere.

The studies conducted at national and international level make a useful start, but much additional information is needed. Some of the major questions that need answering are:

- To what extent are claims of the effectiveness of BPC agents valid under farmer conditions? Do plant attributes vary according to different ecological conditions? Does the Indian neem, for instance, behave the same as the Pakistani neem or the Sudanese neem? Does age of tree make a difference? Are the younger leaves more effective or the older ones? In the case of neem, the fact that researchers are getting mixed results suggests that one or more of these ecological/genotypic factors is operating.
- How do farmers cope with the variable composition of a plant's active constituents? What are their expectations of the level of crop protection afforded by the use of BPC agents?
- How should materials be prepared and applied? What rates of application need to be used? How does the use of botanical materials compare with synthetic pesticides economically?
- Can farmers collect plant parts in the quantities needed? Can facilities be set up at village level to extract the BPC agent from these plant parts? What is the minimum economically viable size of such a unit? What are the costs and the training requirements?
- How safe are the plant extracts for humans, domesticated animals and wildlife? What would be the environmental consequences of introducing promising pest control species into new areas?

Clearly, an enormous amount of data needs to be gathered. But we should not be daunted. A start has already been made. Building on traditional practices, studies under controlled laboratory conditions have been initiated in several countries over the past two to three decades. The results obtained are frequently impressive and, in the case of neem, convincing. Now we need to go further, entering the phase of large-scale field testing and socio-economic analysis for products whose effectiveness is already known, while expanding the range of species under basic laboratory research and field testing.

## Methodology

In biopesticides' development as in other fields, research needs to be firmly rooted in farmers' needs. As the case study from Thailand shows, a participatory approach to the diagnosis of problems and the design of solutions is a useful starting point. However, inputs from formal research and development services will be necessary if real improvements in crop protection are to be achieved. Better access to such services is essential if resource-poor farmers are to benefit.

Figure 4.4 shows the process of technology development used by the NGO in the Thai case study. Several steps can be distinguished. First, in a spirit of open enquiry (without a previously determined technological agenda) field workers approached farmers to discuss their pest control problems with them. The project found that most farmers were well aware of the causes of their problems, but lacked ideas on how to solve them. TREE staff therefore proposed neem-based biopesticides and arranged

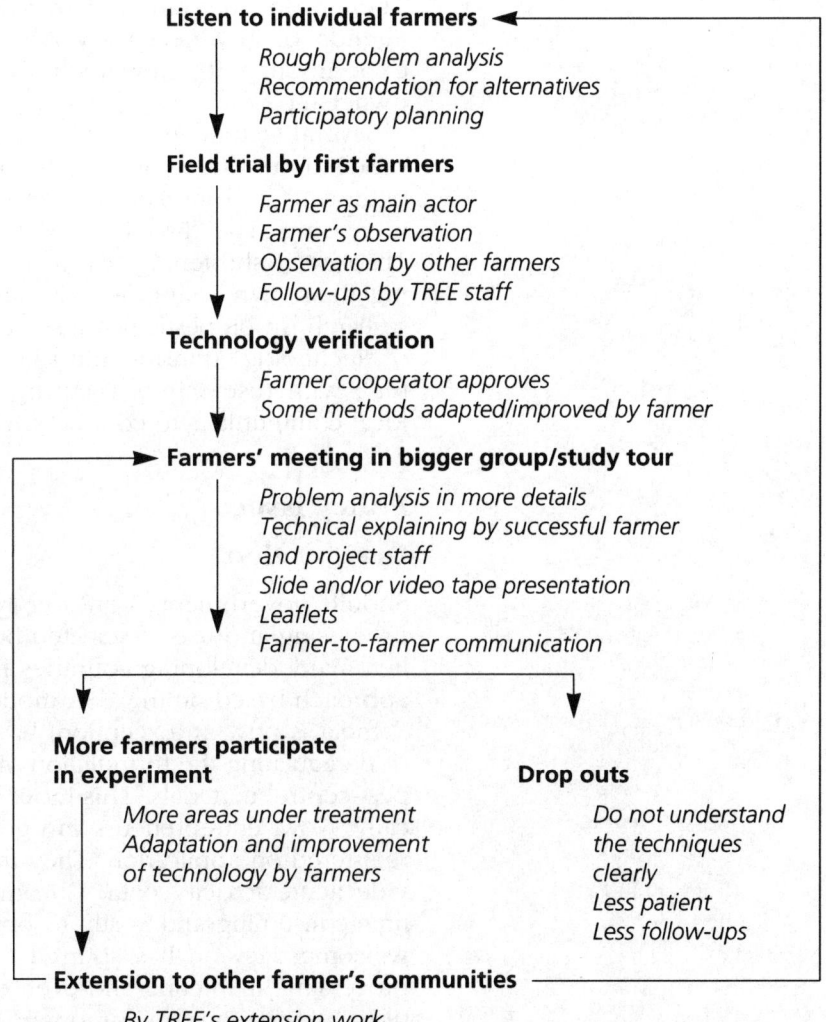

**Figure 4.4**   Process of technology development used in TREE's research.

visits to other farmers with experience in their use. Following these visits, three farmers started testing the new plant protection method in small areas of their fields. On the basis of their observations, farmers and field staff together adapted the technology to suit local conditions. Farmer-to-farmer communication, mediated through TREE and other NGOs, then helped the new technology spread elsewhere.

This is a common sense approach, little different from those used elsewhere by successful practitioners in both the NGO movement and the formal research system. Its importance lies in the fact that farmers' observations and experiences guide the research process and provide the basis for technology adaptation. Their problems, rather than scientists' interests, determine the agenda of any necessary laboratory-based research. As they experiment with new technology, farmers gradually assume ownership of it.

Several factors are quickening the pace of change in traditional production systems in developing countries. The rate at which pathogens and insects become resistant to synthetic pesticides is increasing. The time it takes for "new" knowledge to become obsolete is shortening. Farmers must become expert at developing their own solutions, without waiting for those from formal research. In biopesticides research, as in other fields, the concept of "technology transfer" must be replaced by "methodology transfer", with researchers focusing more on ways of empowering local communities to conduct their own research.

## Policy issues

### Deregulation

Should governments consider relaxing the laws governing the processing and use of selected botanical pesticides? As noted earlier, many developing countries have adopted a highly regulatory approach based on the USA model.

India is now an exception, having recently taken the bold step of deregulating the formulation, marketing and use of neem-based pest-control materials. This move has already borne fruit economically. Now, entrepreneurs can get neem products for pest control registered on application. They then have up to two years to provide acute toxicity data. Chronic toxicity data, which are very time-consuming and costly to obtain, are no longer needed. This welcome decision has spurred many local entrepreneurs to start processing, marketing and even exporting such products. In addition, at least one major organic farming and food processing enterprise in Japan, having more than a million customers, is currently experimenting with crude extracts with a view to marketing similar products in that country (Ahmed, 1994).

On the other hand, deregulation also carries certain risks. As the Thai case study showed, in the absence of effective government controls unscrupulous operators selling inferior products may wreak havoc in what should be a profitable market. Countries must find a middle way between the two extremes of rigid control and total deregulation.

## Effectiveness

A significant milestone was passed recently when two neem-based formulations received EPA clearance for pest-control use. These are now being produced and marketed by US companies (Grace Chemicals and AgriDyne). However, both formulations are based on only one active ingredient (azadirachtin) refined out of the neem "soup" containing 12 to 15 such ingredients. The range of pests controlled is therefore relatively small. Even more worrying is the prospect that pests may in time develop resistance to this chemical. If they do so, there is the added danger that they may also exhibit resistance to the same chemical found in the crude extract, thereby reducing significantly the range of pests that can be controlled by traditional neem users.

These manufactured products are no longer very different from synthetic pesticides, which are also usually based on only one active ingredient. In contrast, the crude extracts of most plant species contain more than one active ingredient, as demonstrated by the majority of the 12 promising plant species described above.

Neem cake being sold by a pesticide dealer in Karnataka, India. (*Saleem Ahmed*).

## Appropriation

The "Who gains?" issue also needs to be examined. This issue is again best illustrated by recent events in India, where a movement is currently gaining momentum to prevent the above-mentioned US companies from establishing large-scale neem oil production facilities. The movement's members argue that an important natural resource will be taken away from the country at a very low price, to be made into pesticides and sold back to the country at a much higher price.

The prices the companies plan to charge are indeed high, probably two to five times those of the equivalent synthetic pesticides. They are based on the willingness of US consumers to pay a premium for safe pest-control formulations. These prices should come down as competitors enter the market with similar products.

Studies are needed to assess how this emotive and potentially explosive issue, which has all the makings of a typical North-South conflict, can be turned into an opportunity for cooperation rather than confrontation. The South's bargaining position is somewhat better than usual, since its resources will be used to tackle an important environmental problem that is, for once, more acute in the developed than in the developing countries. Perhaps a two-tier pricing system, one price for US users and another for the developing countries, is the answer.

## Biodiversity

Another important policy issue is the possible indiscriminate harvesting of botanical materials to meet a surge in demand. This could lead to environmental degradation and the loss of biodiversity, as is already occurring in the case of neem in parts of South Asia. From this point of view it is important that BPC agents are not seen as too successful, until adequate arrangements have been made for their increased production.

Governments will need to consider policies to encourage the widespread planting of promising species, preferably through non-financial incentives. Access to an assured market and the provision of processing facilities spring to mind, as well as improved extension efforts.

## Extension

At the extension stage, governments will need to launch campaigns to promote the use of selected pest-control materials. Issues such as the lack of immediate knock-down or total kill need to be addressed through information and education programmes. Efforts will be needed to change the image of BPC agents, so that these become fashionable rather than being seen as backward. It will also be important to consider how to assist farmers who might be interested in trying out the use of different preparations; how to motivate them to experiment and to learn from other farming groups.

Clearly, no single strategy will be universally applicable, given varying ecological and socio-economic conditions. Here, the concept of the FAO, of monitoring pest populations and varying the control measures to suit changing crop and climatic conditions, will come in useful. In some cases preventive measures may be desirable, while in others curative ones will be more appropriate.

# Conclusion

The development and commercialisation of low-cost, safe and effective BPC agents should be welcomed and promoted by policy makers, scientists, environmentalists and others seeking to improve rural life in developing countries. The materials developed need not match the effectiveness of synthetic pesticides, even partial pest control would be better than no pest control at all, the prevalent situation in many developing countries. Underlying these efforts will be a new concept of what constitutes effective pest control; not dead bugs, but sustained increases in crop yields. The aim will be to persuade all farmers to grow a safe botanical pesticide in their backyard.

Lastly, more basic research, including plant exploration, is needed. With an estimated 2400 plant species worldwide having pest-control properties (Grainge and Ahmed, 1988), who knows how many other "unsungwelcome datacomp heroes" are out there awaiting discovery and utilisation? We will not know until we are willing to give them a chance.

## References

Achundow, A.C. 1889/1968, 'Die pharmakologischen Grundsätze des Abu Mansur Muwaffak bin Ali Harawi', in Kobert (ed), Historische Studien aus dem Pharmakologischen Institut der Universität Dorpat, Halle, 1889–96 (1889), compiled under: Historische Studien zur Pharmakologie der Griechen, Römer und Araber, Zentralantiquariat der Leipzig, pp. 138–446

Agrios, G.N. 1978, *Plant Pathology*, Academic Press, New York

Ahmed, S. 1994, 'Teikei: A farming-consumer alliance succeeds in Japan', *Ceres*, July–August. FAO, Rome

Ahmed, S. 1984, 'The use of plant species for pest control in South Asia: Some observations', working paper, East-West Center, Honolulu

Ahmed, S. 1988, 'Farmer pest-control practices in India: Some observations', paper presented at the final workshop of the IRRI-ADB-EWC Project on botanical pest control in rice-based cropping systems, held at IRRI, 12–16 December 1988, Manila, Philippines

Ahmed, S. and Grainge, M. 1986, 'Potential of the neem tree (*Azadirachta indica*) for pest control and rural development', *Econ. Bot.* 40 (2): 201–209

Ahmed, S., Milan, P. and Sultan, Z. 1988, 'Use of plant materials for pest control by farmers in Leyte and Samar, Philippines: Results of a survey', paper presented at the final workshop of the IRRI-ADB-EWC Project on botanical pest control in rice-based cropping systems, held at IRRI, 12–16 December 1988, Manila, Philippines

ATCRD. 1984, 'Inventory of plants with insecticidal properties in the Philippines', progress report, ATCRD, Manila, Philippines

Berger, A. 1994, 'Using natural pesticides: Current and future perspectives', a report for the Plant Protection Improvement Programme in Botswana, Zambia and Tanzania, Department of Entomology, Swedish University of Agricultural Sciences, Uppsala

Chejew, V., Hiranmusaphon, S. and Srihakam, S. 1988, 'Alternatives to chemical pest control', ATA and HDF, Bangkok, Thailand

Chopra, R.N., Bhadwar, R.L. and Ghosh, S. 1949, 'Poisonous plants of India', *Scientific Monograph* No. 17, Indian Council of Agricultural Research, New Delhi, pp. 10–12

Dinham, B. 1993, *The Pesticide Hazard: A Global Health and Environment Audit*, Zed Books, London, England

Duarte, B.O. and Franciosi, T.R. 1976, 'Recent advances in the propagation of some tropical and sub-tropical fruit species in Peru', *Acta Horticulturae* 57: 15–20

Elwell, H. and Maas, A. 1995, *Natural Pest and Disease Control*, Natural Farming Network, Harare, Zimbabwe

Escano, C.R., Quintana, E.G. and Talan, I.F. 1988, 'Survey on the use of plant materials for pest control in selected areas of the Philippines', paper presented at the final workshop of the IRRI-ADB-EWC Project on botanical pest control in rice-based cropping systems, held at IRRI, 12–16 December 1988, Manila, Philippines

Feinstein, L. 1952, 'Insecticides from plants', in *Insects: Yearbook of Agriculture*, USDA, Washington DC, USA, pp. 222–229

Frear, D.E.H. 1943, *Chemistry of Insecticides and Fungicides*, Van Nostrand, New York

Giles, P.H. 1964, 'The storage of cereals by farmers in northern Nigeria', *Trop. Agric.* 42: 197–212

Grainge, M. and Ahmed, S. 1988, *Handbook of Plants with Pest-Control Properties*, Wiley Interscience, New York

IITA, 1990. Annual Report 1989/90, International Institute of Tropical Agriculture Ibadan, Nigeria

Jacobson, M. 1975, 'Insecticides from plants: A review of the literature, 1954–1971', *Agriculture Handbook* No. 461, United States Department of Agriculture, Washington DC, USA

Lianchamroon, V. 1994, 'The experience of "Natural Plants", a registered company producing neem-based products', paper presented at the International Neem Conference in the Dominican Republic

Librero, A.R., Fabro, R.M. and Limbo, R.C. 1988, 'Socio-economic aspects of botanical pest control technology in selected areas of the Philippines', paper presented at the final workshop of the IRRI-ADB-EWC Project on botanical pest control in rice-based cropping systems, held at IRRI, 12–16 December 1988, Manila, Philippines

Mackauer, M., Ehler, L.E. and Roland, J. (eds), 1990, *Critical Issues in Biological Control*, Intercept, Andover, UK

McEwen, F.L. 1978, 'Food production: The challenge of pesticides', *BioScience* 28: 773–777

Munos, L.G. 1985, *Plantas con Propriedades Pesticidas Utilizadas en la Protección Vegetal*, CEMAT, Guatemala City

Nelson, R.H. (ed), 1975, *Pyrethrum Flowers*, McLaughlin Gormley King, Minneapolis, USA

Pagon, C. and Morris, M.P. 1953, 'A comparison of the toxicity of the mammey seed extracts and rotenone', *J. Econ. Entomol.* 46 (6): 1092–1093

Rodriguez, C.H., Lagunes, A.T. and Arenas, C.L. 1984, *Extractos Acuosos y Poivos Vegetales con Propriedades Insecticidas*, Consejo Nacional de Ciencia y Tecnologia, Colegio de Postgraduados, Universidad Autonoma Chapingo, Mexico

Secoy, D.M. and Smith, A.E.. 1983, 'Use of plants in control of agricultural and domestic pests', *Econ. Bot.* 37: 28–57

Shiva, W. 1988, *Staying Alive: Women, Ecology, and Survival in India*, Kali for Women, New Delhi

Shrimpf, B. and Pag, H. 1993, 'Evaluation report on the alternative plant protection programme in Vietnam', Bread for the World, Frankfurt, Germany

Srithong, P., Lianchamroon, W. and Stoll, G. 1994, 'Development of neembased plant protection practices: A PTD experience from Suphanburi, Thailand', unpublished paper

Stoll, G. 1992, *Natural Crop Protection Based on Local Farm Resources in the Tropics and Subtropics*, Josef Margraf, Weikersheim, Germany

Thiam, A. and Ducommun, G. 1993, 'Protéction naturelle des végétaux en Afrique', proceedings of a workshop organised by ENDA-PRONAT in collaboration with AGRECOL (Switzerland) at Madesahel, Senegal, 21–26 October 1991, Dakar: ENDA-Editions

Yang, R.Z. and Tang, C.S. 1988, 'Plants used for pest control in China: A literature review', *Econ. Bot.* 42 (3): 376–406

# 5 Food Processing

*Andrew Jones, Vishaka Hidellage, Helen Wedgewood,*
*Helen Appleton and Mike Battcock*

## Introduction

Processing food makes it more palatable, nutritious, varied and stable for storage purposes. It also adds value to basic foodstuffs and enables them to be sold to a larger market, increasing incomes and creating employment. These benefits apply in both developed and developing countries, but because of the larger share of food production and exports in the gross domestic product of developing countries the technologies associated with food processing are of special importance there. Up to 80% of the population in developing countries live in rural areas and use these technologies in their daily lives.

Many traditional food processing practices used in developing countries can be classified as biotechnologies. The key process involved is fermentation. Almost any foodstuff can be subjected to a fermentation process of some kind, with the result that the range of fermented food products available around the world is highly diverse. A few products, including such staples as cheese, bread, wine and beer, are now mass-produced on an industrial scale, but many more are still made exclusively on a small scale at village or household level. Even where industrial products have come onto the market, their traditional counterparts remain in production in many rural areas. Traditional processing practices may be highly sophisticated, although the equipment used is usually simple: Sudanese women allegedly carry out a complicated 40-stage fermentation process using little more than a gourd (Dirar, 1993).

Traditional biotechnology is not something static. Its products and practices have been adapted over the generations. Some have fallen by the wayside; others have, thus far, stood the test of time. Adaptation is likely to continue and intensify in response to the changing demands of the market-place. The environment in which traditional biotechnology must survive is becoming more competitive, as subsistence economies are transformed into market-oriented ones. Rising incomes and the emergence of a new middle class are creating a demand for new products. In response, the raw materials available for processing are themselves changing. Traditional food grains, such as sorghum and millet, are giving way to rice and wheat, which are faster to prepare and are considered to be more refined in taste or texture.

There is also a general shift in demand away from cereals towards vegetables and livestock products (meat and milk).

In some cases the demand for new products can be met by adapting traditional biotechnologies or developing new ones. For example, cassava is now used to make a range of new products, including bread and animal feeds, besides traditional foods such as *gari* and *fufu* in West and Central Africa. In other cases, traditional processing activities, and with them the livelihoods of rural people, find themselves in decline as they face increasing competition from the introduction of Western foods. These foods may either be imported or mass-produced locally. In the former case, they may depress prices for home-grown products. In the latter, they can sometimes destroy more jobs than they create.

Most of the indigenous knowledge and skills associated with food processing in rural households are held by women. This makes the issue of the visibility of women important in the development of rural peoples' biotechnology.

# Fermented foods

## Benefits

Table 5.1 summarises the advantages of including fermented foods in the diet. There are, of course, some disadvantages too, associated mainly with the production and consumption of alcohol.

For poor people, traditional food processing is closely linked to survival. Below we outline its specific roles in their survival strategies.

### Food storage

Food processing allows rural people to store food, and so to bridge the hungry gap or lean season, the time of year when food is scarce before the new harvest comes in. It also provides a food reserve against a year when crops fail altogether. It also enables farmers to avoid having to sell surpluses at low prices during the post-harvest glut. Sudan provides an example of a country in which most processed food products seem to have been developed to overcome food shortages caused by drought (Dirar, 1993).

Food usually needs to be preserved before storage. Typical preservation techniques include drying and fermenting, to produce foods such as *gundruk* (pickled vegetable leaves in Nepal), *iru* (a dried paste made from locust beans in Nigeria and other West African countries) and *gari* (fermented and dried cassava, also produced in Nigeria and elsewhere in West Africa).

### Food quantity and variety

Food processing increases the quantity and variety of commodities available for human consumption. Striking examples of this

TABLE 5.1   Benefits of fermented foods

| Contribution to the economy of food winning | Protective value | Psychological and social value |
| --- | --- | --- |
| Preserves perishable raw material at low cost | Reduces or destroys toxic, undesirable or anti-digestive components of the raw material | Improves appearance and flavour, often imparting a meat-like flavour |
| Salvages waste otherwise not usable as food | Increases the range of raw materials available as food | Stimulates appetite and can be used as condiments |
| Reduces cooking time, hence reducing demand on food winning time by reducing fuel demand | Adds positive antibiotic components, destroys harmful biota | Imparts texture, fibre and bite/chewiness Makes the product enjoyable |
| Enhances nutritional value by improving digestibility, protein value and vitamin content | Protects against re-infection or infestation | Methods used can be inexpensive Techniques are simple and well understood Products are well established and acceptable |

Source:  Stanton W.R. in Wood B.J. *Microbiology of Fermented Foods* Vol 2, (1985).

are the immense ranges of cheeses and wines available in Europe. Another example is the processing of cassava, which is detoxified by grating and washing. If the detoxification process were unknown, most cassava roots would be unavailable for human consumption. Instead, this drought-resistant crop that is ideally suited to poor soils has become a staple food for millions of people in West Africa and elsewhere. In Sudan, a fermentation process enables the leaves of the otherwise toxic legume *Cassia obtusifolia* to be turned into a valuable dish used to replace meat in the diets of poor people. These and many other examples illustrate the value of the knowledge gained through indigenous experimentation and innovation (Dirar, 1992).

## Quality of diet

Food processing can improve the quality of rural people's diets. This is especially the case for foods not easily stored in their initial form. For example, the fermentation of milk products enables the protein content of human diets to be sustained over long periods. Dairy processing is an ancient human skill; knowledge of cheese and butter making is recorded in ancient Indian writings dating back some 3000 years (Aguhob and Axtell, 1994).

A traditional household technology throughout much of Africa, the fermentation of cereals improves not merely their keeping quality but also their digestibility. This is particularly important for young children (see Box 5.1)

## Income generation

Food processing also contributes to survival by adding value to primary commodities and so increasing household incomes. This income-earning potential is often realised by women, who make food products for sale at roadsides or local markets.

Traditional food products, including those created using biotechnology, can play an important part in the economies of developing countries. In the Bangladesh town of Manikganj, which has a population of 38 000 people, traditional street food businesses such as pickles and yoghurt have a turnover of US$ 2 million a year and employ up to six per cent of the workforce. The average wage of workers in the street food sector is three times the average agricultural wage (Tinker and Cohen, 1986).

The sector is ideally suited to the needs of poor people, and poor women in particular. People setting up a small food processing business do not need vast amounts of money. Most of the equipment required is available in the domestic kitchen, and small quantities of locally available raw materials are all that is needed to get started. Many women have been trained by their mothers to make these products. They can often do so at the same time as looking after children and without leaving the homestead. There is a fast return to initial investment: a new business can buy the raw materials, make the product, sell it for a profit and receive payment in cash, all in one day.

---

**Box 5.1**
***A way to reduce child mortality***

During weaning, Africa's young children are particularly vulnerable to malnutrition. Food may be scarce in absolute terms, reducing the amount their mothers can provide. If the cereals used as weaning foods are prepared simply as a liquid gruel, they may not be very nourishing or digestible. Poor digestibility can cause diarrhoea, weakening the child and reducing its intake still further. And food is often contaminated during preparation, increasing the likelihood of infections.

Fermentation helps solve these problems. Lactic acid producing bacteria lower the pH of foods, inhibiting the growth of other bacteria that could cause decomposition and spoilage. They also contain proteolytic enzymes capable of degrading complex proteins into simple proteins, peptides and amino acids. This aids digestibility, especially in coarse-grained cereals such as sorghum and millet. During fermentation phytates and tannins are degraded, increasing the available iron content of the diet and so helping to prevent anaemia.

In a study conducted in two Tanzanian villages, children receiving fermented gruels had a 33% lower incidence of diarrhoea than those fed on unfermented gruels. Fermented cereal products are already widely produced in Africa (Table 5.2). Spreading the practice of fermentation offers perhaps the simplest and most economical means of improving the nutrition and health of the continent's children.

Source: Svanberg (1992).

TABLE 5.2   Fermented cereal foods in Africa

| Product | Cereal | Microorganism involved | Nature of use | Regions |
|---------|--------|------------------------|---------------|---------|
| Uji | Maize, sorghum, millet | Lactobacillus plantarum | Liquid drink for infants and young children | Tanzania, Uganda, Kenya |
| Kenkey | Maize | Aspergillus, Streptococcus, Lactobacillus | Thick dough for adults and children | Ghana |
| Nasha | Sorghum, millet | Streptococcus, Lactobacillus, Saccharomyces | Liquid drink especially for sick children | Sudan |
| Mahewu | Maize | Lactic acid bacteria | Liquid drink for adults and children | South Africa |
| Ogi | Maize, sorghum | Lactic bacteria, yeasts and moulds | Paste for infants and young children | Nigeria, West Africa |
| Bogobe | Sorghum | Not shown | Soft porridge | Botswana |
| Injera | Tef, sorghum | Candida | Pancake, Staple food | Ethiopia |

Source: Svanberg (1992).

## Processes

Fermentation is a complex group of processes. The basic raw material (which can be a cereal, legume, fruit, vegetable, milk, meat or fish) becomes the substrate or medium for the controlled growth of specially selected micro-organisms.

The factors determining selection and growth include treatments such as chopping and boiling, added ingredients such as salt, and environmental factors such as temperature. The right micro-organisms flourish only when the right combinations of nutrients, temperature, moisture and acidity/alkalinity occur. They are then able to suppress other organisms which might cause food poisoning or spoil the process. Managing fermentation so as to arrive at the desired end-product is thus a highly skilled business.

As they grow, the micro-organisms produce their own by-products, such as acids or antibiotics. Some of these products are enzymes, catalysing the conversion of certain components of the original raw material to other products. For example, starches are converted to sugars, sugars to alcohols or alcohols to acids. With these chemical transformations come changes in the texture and palatability of the raw material. For example, liquid milk becomes solid (yoghurt, cheese), individual legume seeds coagulate to form a solid mass with a meat-like texture (*tempeh*) and solids turn into liquids (fish sauces). Metabolites are another important group of by-products. These play a beneficial role in enhancing texture and flavour, and also in preserving the end-product.

The main groups of micro-organisms involved in fermentation are discussed below.

- Lactic acid bacteria (*Lactobacillus* spp). These assist in the fermentation of all dairy products, and are often an ingredient in other fermented products made from cereals and vegetables. Acid foods are less susceptible to spoilage than neutral or alkaline foods and hence lactic acids help preserve foods. They also cause coagulation, as in the case of curds and yoghurts.
- Acetic acid bacteria (*Acetobacter* spp). These convert alcohol to acetic acid. They are responsible for the production of vinegar from wines and beer.
- Other bacteria. These include the propionic bacteria, employed in the production of cheeses such as emmental and gruyère, and the *Bacillus* species, found in *dawadawa* (see page 88).
- Yeasts. These typically carry out alcoholic fermentations, converting sugars to alcohol as in the production of beer, wine and bread. They are also used in mixed fermentations with lactic acid bacteria, producing a mildly acidic or alcoholic flavour that is often very popular with consumers. Typical yeast-lactic fermentations include sour dough bread and milk products such as *koumiss* (from East Europe). It is thought that some cereal and root products also involve a yeast as well as a lactic fermentation (Wood and Hodge, 1985). *Ogi*, a Nigerian porridge, is one of them (see page 86).
- Moulds. Products such as soy sauce, *tempeh* (fermented legume seeds from South-East Asia), and blue-veined cheeses are the result of fermentations by moulds belonging to the *Aspergillus, Rhizopus* and *Penicillium* genera.

## Products

There are hundreds of traditionally fermented foods. We describe here a selection that illustrates the range and variety of products in different regions of the world. Some the reader will almost certainly recognise; others are less well known. Our selection is drawn partly from the results of the ETC contest (see Introduction) and partly from recent publications that have provided descriptions of locally produced fermented foods (Wood, 1985; Gaden, 1992; CFTRI, 1986).

## Cereals

The majority of traditional fermented foods are made from cereals. All the major cereals can be processed using traditional biotechnology. Major products include porridges, bread and alcoholic drinks.

Lucy, Anna and Hannah Wanjiku demonstrate how they make fermented porridge or *uchuru*. After adding water, grains from dry maize mixed with millet are soaked in water for 12 hours. After pounding, sieving and filtration it is left to ferment in gourds (*njanja*) for 48 hours. Portions of the fermented liquid are added to warm water and before serving it as *uchuru*, sugar is added to taste.
(*Kees Manintveld*).

### Porridges

These are made from a wide range of cereal grains. In Nigeria, *ogi*, made from maize, sorghum or millet, is an important weaning food as well as a breakfast dish (Adeyemi, 1993). The cereal is steeped in water for one or two days to soften the grain. The grains are then milled and water is added to extract the starch in solution. The starch is left to separate from the water and then allowed to ferment naturally. This can take two to five days. The paste or slurry is diluted with water, boiled down to the consistency of porridge and then sweetened before consumption (Figure 5.2) (Mwangi and Mwangi, 1993).

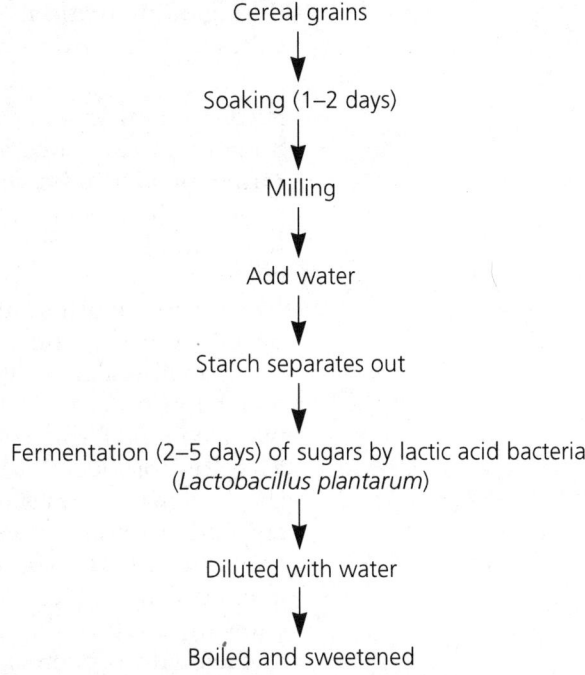

Cereal grains

↓

Soaking (1–2 days)

↓

Milling

↓

Add water

↓

Starch separates out

↓

Fermentation (2–5 days) of sugars by lactic acid bacteria
(*Lactobacillus plantarum*)

↓

Diluted with water

↓

Boiled and sweetened

**Figure 5.2**    Flow diagram for the production of *ogi*.

*Bread*

The conversion of cereal into bread is one of the world's most
ancient and varied food processing technologies. Countless kinds
of bread have been made down the ages, usually by women.
Most modern recipes require the addition of yeast (*Saccharomyces
cerevisiae*), which ferments to produce alcohol and carbon dioxide, causing the rising or leavening of the dough. However, many
traditional breads depend on a natural combination of yeasts and
lactic acid bacteria to effect the leavening process and impart a
characteristic sour flavour.

A range of grains, sometimes mixed with legumes, is traditionally used to make these sour dough breads. For example *kisra*, in
Sudan, is made from sorghum, while *puto*, in the Philippines, is a
fermented rice cake. *Idli*, a product widely consumed in India,
consists of a mixture of rice and black gram beans.

The basic process for sour dough fermented breads is to mix the
flour with water, salt and a starter. The latter is usually a small
amount of fully fermented dough from a previous batch. The new
dough is allowed to ferment in a warm place before cooking.
Cooking techniques vary: *kisra* is baked in thin sheets like a pancake, whereas *idli* is steamed (Steinkraus, 1992). Puto is made
from older rice grains which are ground and allowed to ferment
with the addition of sugar. A feature of this process is the addition
of lye (sodium hydroxide), which reduces the acidity produced

during fermentation. The rice flour slurry is then shaped into cakes and steamed (Sanchez, 1986).

### Alcoholic drinks

Sorghum, maize, rice and other cereals are used to make a wide range of alcoholic drinks. Those produced on an industrial scale, such as beers, use yeasts to ferment the grains. Most traditional drinks involve the action of both yeasts and lactic acid bacteria.

Two cases illustrate the different types of beer that can be produced from sorghum. *Merissa* is a Sudanese beer, still produced on a small scale at village level. Producers use a three-stage fermentation process, in which a lactic fermentation is followed by two alcoholic fermentations. *Kaffir* is from southern Africa, where it is now produced on an industrial scale. The brewing process involves malting, mashing, souring, conversion of starches to sugars, and alcoholic fermentation (Odunfa, 1985).

*Chicha*, a traditional drink of South America, is made from fermented maize or cassava. Producers trigger fermentation by first chewing some of the maize, a process which enables an enzyme in their saliva to break down the starches to sugars. These then ferment to alcohol. Other beers produced from maize are popular in Africa, including *busaa, malawa* and *pito* from Kenya, Uganda and Nigeria respectively.

In Nepal, *jand* is a beer type product prepared from rice. In Japan, *saké* or rice wine is an immensely popular drink prepared from an alcoholic fermentation of rice pre-treated with *koji* (a starter taken from a previous batch) to convert the starches to sugars. *Saké* is easy to make on a small scale and the product has been widely exported all over Asia.

## Legumes, oilseeds and nuts

Fermented legumes, oilseeds and nuts can acquire a strong flavour and are widely used as condiments.

*Iru* or *dawadawa* is prepared from the locust bean, more commonly known as carob or the alternative chocolate. It is used in many African dishes, especially in Nigeria. Its high protein content makes it a valuable component in the diets of poor families. Traditional *iru* processing is still carried out at domestic level, but the product's popularity has led to the development of a modern version similar to a stock cube, produced industrially in Nigeria (Odunfa, 1985).

Traditionally, *iru* is made by boiling dried locust beans for up to 24 hours. The seed coats are removed by adding ash, after which the seeds are again boiled for about two hours. While the seeds are still very hot they are drained, transferred to a basket and covered with leaves. This helps to retain heat and create a humid atmosphere. The beans are allowed to ferment for two to three days. The leaves are then removed and the seeds are crushed to a paste. Various shapes can be moulded from the paste, and these are sun dried (Figure 5.3).

*Tempeh* is a meat-like product made by soaking, dehulling and partially cooking soya beans, then adding *Rhizopus* mould. The mixture is kept in a warm place (30–33°C) for one to three days. The mould growth helps to improve the digestibility of the legumes used and also binds the beans together into a solid form, which can then be cut into small blocks or thinly sliced and used in place of meat (Steinkraus, 1985).

Home-made soy sauce is made from small balls of cooked and mashed soy beans which serve as the substrate for the mould (*Aspergillus oryzae*). After ageing, the balls are placed in brine to promote fermentation. Eventually they dissolve, forming soy sauce.

Other fermented products in this category include *ogiri*, made from castor seed, *ugba*, from oil bean (Ezedinma and Igbinnosa, 1993) and *koji*, from soybean (and wheat), which is the starter culture for soy sauce. *Ogiri* is made throughout West Africa, and particularly in Sierra Leone and Nigeria, while *ugba* is common in Nigeria and *koji* in China. Also in Nigeria and other West African countries, palm wine and gin are widely produced at local level (Iji, 1993).

## Roots

Roots and tubers are important constituents in the diets of millions of people. Traditional biotechnologies have been developed to help preserve them and add variety to the diet. However, relatively few root crops are subject to processes involving fermentation, the main one in developing countries being cassava.

The chief role of fermentation in the case of cassava is to aid preservation and introduce variety into cassava-based diets

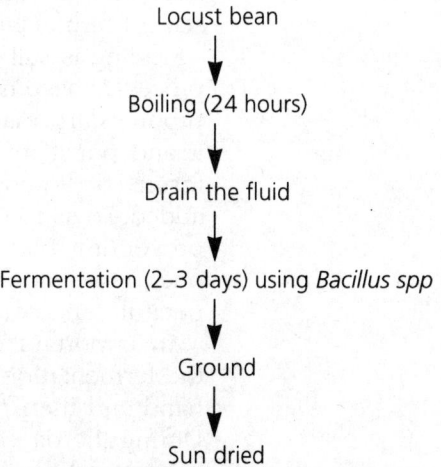

**Figure 5.3**   Flow diagram for the production of *iru*.

(Sokari, 1992; Martin, 1993). The two main traditional products are *gari* and *fufu*. In the former, washed, grated and pressed cassava is allowed to ferment in cloth sacks, after which the coarse fibres are sieved out and the remaining pulp is fried. In the latter, fermentation is achieved by submerging cut pieces of washed and peeled cassava in water. The fermented pieces are then sieved, pressed and made into small balls, which are boiled and mashed into a dough.

Another fermented cassava product is sour starch, made in several countries of South America (Thro, 1993). In a process similar to that for *ogi*, the starch is fermented in water until the requisite degree of souring has been achieved. The elastic properties of the fermented starch make it suitable for making bread.

## Vegetables

Fermentation is a traditional way of preserving surpluses of vegetables that cannot be consumed immediately after harvest and would otherwise have to be sold at low prices. Fermented vegetables serve several purposes. Highly flavoured products such as pickles and *gundruk* are used as condiments. Sauerkraut (pickled cabbage) is eaten as either a main or a side dish in Germany and Central Europe. *Kawal*, made in Sudan from the leaves of the otherwise toxic legume *Cassia obtusifolia*, is used to replace meat in the diets of poor people, while fermented cassava leaves are made into a dish with meat or fish to accompany the staple diet of cassava tuber.

Pickles are made by storing prepared vegetables in a weak brine solution, or sometimes by dry salting them. Either method provides a suitable environment in which lactic acid bacteria can grow, imparting a characteristic acid flavour. Nearly all vegetables can be pickled, the ones most commonly processed in this way being cucumber and cabbage. Some pickled products have the additional flavour of dill, which can be added to the brine at the start of fermentation.

Pickling is still carried out at domestic level. However, industrial processes have now been developed for many of the more popular products. To make sauerkraut, producers shred cabbage, dry salt it and put it into barrels covered with plastic sheets. A brine is formed as liquid is drawn out of the cabbage. More brine may be added, so as to ensure that the cabbage is completely submerged, preventing blackening. Fermentation is usually complete after one to two months, although the period required varies with temperature and concentration of the brine (Vaughn, 1985).

An important condiment in Nepal, *gundruk* is obtained from the fermentation of leafy vegetables. The leaves are pounded by hand and then placed in an earthenware pot, with straw on top. During the day the pot is left outside in the sun; in the evenings it is placed near a heat source. Warm water is periodically added to keep the pot brim-full and to maintain the temperature of the fermenting mass. After fermentation the product is spread out on clean mats to dry in the sun (Karki, 1986).

*Kawal* is a similar product to *gundruk*. The leaves of *Cassia obtusifolia* are pounded to a paste and packed into an earthenware pot, which is buried in the ground up to its neck. Sorghum leaves are placed on top of the paste and the whole pot is sealed. Every three days the pot is taken out of the ground and its contents are remixed, before the pot is replaced in the ground as before. Fermentation is complete after about 15 days. The paste is rolled into small balls, which are then dried in the sun (Harper and Collins, 1992).

In Zaire a complex process is used to make a product called *ombolo wa koba*. Harvested cassava leaves are allowed to wilt and turn black, which takes about three to four days. Meanwhile, *koba* (ash) is prepared by burning dried banana skins and palm flowers. The blackened cassava leaves are chopped up and placed in a pot of boiling water for about one hour. A water-soluble extract of the ash is obtained by placing the ash in a strainer and pouring water through it. This extract is added to the cassava leaves after boiling. The extract is alkaline and hence neutralises any cyanhydric acid liberated when the leaves are chopped up and not totally destroyed during boiling. Also added after boiling are salt and dried fish or meat. The mixture is allowed to cool a little, before acid palm oil is added. This reacts with the excess alkali, which it neutralises. The end product is traditionally served with boiled cassava and plantain bananas (Menea and Bishosha, 1993).

## Fruits

Fermented products from fruits are mostly alcoholic drinks and vinegars. Some are pickled, however – olive being an example.

Virtually any fruit can be processed into an alcoholic drink. The process is well known and, in the case of wine from grapes, well researched. Essentially, it is an alcoholic fermentation of sugars to yield alcohol and carbon dioxide.

Banana beer is produced widely in the relatively high-rainfall mid- to high-altitude areas of East and Central Africa. It is a rare example of a traditional alcoholic drink which has been the subject of some attention from the formal research system, although much remains unknown (see Box 5.2).

Vinegar, an important condiment and preservative worldwide, is produced by an additional fermentation after the fermentation of sugars to alcohol. It is the acetic acid produced by the fermentation of alcohol (ethanol) which gives the product its characteristic flavour and aroma. A wide range of raw materials can be made into vinegar. Traditional vinegars include palm wine vinegar, reported by Koleoso and Kuboye (1986), and coconut vinegar, produced from the sap of palm flowers. The sap ferments to toddy, to which a vinegar starter (from a previous batch) is added to precipitate the second fermentation.

**Box 5.2**
**Banana beer: A woman's business**

Home to over 50 cultivated varieties, the East African highlands are a secondary centre of diversity for banana. Each variety has a specific end use; either for beer, for steaming, boiling or roasting as cooked food, or for eating raw as a dessert. Banana beer plays a strong part in creating social equity in the region, being an important source of cash for many women in rural areas and particularly for groups such as widows and divorcees, who are often not permitted to keep the income accruing from more modern crops such as coffee.

Women make banana beer from mature, green fruit which they warm before peeling and pressing, using wads of grass to aid the extraction of juice. A kilo of bananas yields around 500 ml of a clear juice with a good aroma. Fermentation takes place out-of-doors, often in the banana orchard, where the fresh juice is transferred to a "banana canoe" fashioned from a hollowed tree-trunk. Freshly roasted and coarsely ground red sorghum or millet is added. During fermentation the juice is covered with freshly cut banana leaves or stem sections, with the grass wads previously used in extraction heaped on top. Fermentation takes between 12 hours and three days, after which beer is decanted, filtered (removing the cereal grains) and cooled. Diluted beer keeps for a week, undiluted for a month or so.

For most traditional fermented products the micro-organisms responsible for fermentation are unknown to scientists. Banana beer, however, is an exception. A study by Munyanganizi (1974) in Zaire and Rwanda revealed that *Saccharomyces cerevissae,* which also initiates vinification in grapes, is the principal causal agent. Many other micro-flora were also found, but there was no trace of *Schizosacchsromyces pombe,* often proposed as the principal microflora associated with African beers and wines (Table 5.3).

Many questions surround other aspects of the process. The effect of the grass wads or other materials used to aid pressing is not known. The role of the added cereals also remains unclear: some authors believe them to contain the starter medium that triggers fermentation, but others question whether the micro-flora could survive prior roasting in sufficient numbers. Other aspects of interest include the special varietal characteristics of juicy banana varieties and the ripening effect of warming before peeling. On these and other topics, formal researchers have much to find out before they can contribute usefully to improving traditional methods.

Source: Davies (1994).

## Milk

Fermentation is one of the oldest, simplest and cheapest methods of preserving milk. Again, many different products are made in different regions of the world. Products vary considerably according to both the process used and the species of animal from which the milk was drawn.

*Dahi* is an Indian fermented milk not unlike yoghurt. It can be made from buffalo or cow's milk (Kumar, 1993). The milk is innoculated with *Streptococcus lactis, S. thermophilus, Lactobacillus bulgaricus, L. plantarum* and various lactose utilising yeasts. It is then kept overnight near an oven. *Dahi* made from buffalo's milk has a higher curd level.

*Ayib* is a type of cottage cheese made in Ethiopia (Ashenafi, 1992). Producers first churn soured milk to produce butter. The buttermilk that is a by-product of this process is then heated gently to precipitate the curd or *dahi*. Eight litres of milk yield about one kilogram of *ayib* (Figure 5.4). *Ayib* is often served as a cool, refreshing accompaniment to the famous spicy national stew dish with soured pancake bread, known as *wat* and *injera*.

TABLE 5.3    Colonies of micro-flora found in banana beer at six sites in Zaire and Rwanda

| Country | Zaire | | | | | Rwanda | Total |
|---|---|---|---|---|---|---|---|
| | (1) | (2) | (3) | (4) | (5) | | |
| Province | Goma | Ruthuru | Kabare | Kalehe | Walungu | Nyundo | |
| Saccharomyces cerevisae | 32 | 5 | 33 | 16 | 14 | 2 | 102 |
| Saccharomyces italicus | 10 | 12 | 5 | 8 | 3 | – | 38 |
| Saccharomyces bayanus | 4 | 2 | 3 | 1 | 3 | – | 13 |
| Saccharomyces oleaginosus | – | 1 | – | – | – | – | 1 |
| Saccharomyces capensis | 2 | 1 | – | – | 1 | – | 4 |
| Saccharomyces montannus | 3 | – | – | – | – | – | 3 |
| Saccharomyces heterogenius | 1 | 1 | – | – | – | 1 | 3 |
| Saccharomyces chevalieri | – | – | – | – | 1 | – | 1 |
| Kluyveromyces fragilis | 2 | – | – | – | – | – | 2 |
| Pichia membranaefaciens | 4 | 2 | 5 | – | 2 | – | 13 |
| Hansenula anomala (var. anomala) | 1 | 2 | 1 | 3 | 1 | – | 8 |
| Kloeckera apiculata | 12 | – | 2 | – | 1 | – | 15 |
| Candida and Torulopsis | 6 | 4 | 5 | 2 | 7 | 1 | 25 |
| Total: | 77 | 30 | 54 | 30 | 33 | 4 | 228 |

Source: Munyanganizi (1975).

*Durkha*, a fermented yak milk from Nepal, is made in a similar way to *ayib*, but with one additional step. The curd obtained from heating the buttermilk is placed in a cloth bag and then squeezed to extract most of the liquid. The pressed curd is left to hang for a few days, after which it is cut into slices and dried (Karki, 1986).

*Koumiss* and *kefir* are two traditional milk products from Eastern Europe. The main ingredient of *kefir* is *kefir* grains, which are recovered from sour milk and contain the micro-organisms required for the fermentation. These include *Saccharomyces kefir, T. kefir, Lactobacillus caucasicus* and *L. casei*. The grains are added to milk and fermentation begins. The end product has a slightly alcoholic flavour. *Kefir* grains can be stored in water for a few days or dried and kept for more than a year.

*Koumiss*, traditionally made from mare's milk, is made by adding *koumiss* from a previous batch to fresh milk held in a skin bag. Fermentation proceeds for up to eight hours, by which time the product has developed an alcoholic flavour (Oberman, 1985).

## Meat and fish

Meat and fish are highly perishable in their fresh state. Fermentation provides a low-cost way of preserving them at the same time as adding variety to the diet.

When fresh, meat and fish have a neutral to alkaline chemical balance. These are conditions under which pathogenic bacteria thrive, increasing the risks of food poisoning. Hence greater precautions must be taken when processing these foods. When properly carried out, fermentation quickly results in increased acidity, inhibiting the growth of poisonous bacteria.

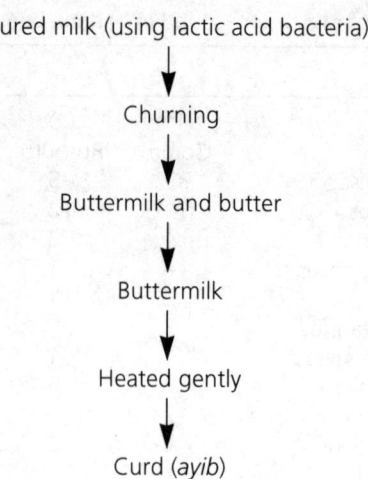

**Figure 5.4**    Flow diagram for the production of *ayib*.

There are three basic types of fermented fish: fish sauces, fish pastes and salted fish (Dirar, 1992; Beddows, 1985). In sauces and pastes, fermentation may bring about considerable changes in texture and flavour.

*Bagoong* is the name given to a fish paste produced in the Philippines. Salt is mixed with whole or ground *Stolephorus*, *Sardinella* or *Decapterus* spp., fish roe and shrimps to a ratio of about one to three. The mixture is then placed in large vats to ferment until it has the desired characteristics. Proteolytic enzymes in the fish break down the tissue. Excess moisture is drained off. The process may take one to two months.

*Patis* is a fish sauce, also from the Philippines. It is made in a similar way to *bagoong*, but the fermentation period lasts for between one and two years. Enzymes in the fish gut react with the proteins to form a liquid, which is collected by draining and filtering the *bagoong* (Winaro, 1986).

Fermented meat products include hams and some sausages. The latter are described in detail by Lücke (1985). Dirar (1992) presents a wide range of fermented meat products from Sudan, including *shermout*, made from fatty strips of meat, and various other products made from the intestines and offal of small ruminants.

## Producers' problems

In this section we summarise the major categories of problem faced by small-scale producers of fermented foods.

### Scarcity

Changing demand is not the only factor inducing changes in the food processing subsector. Other factors include seasonal or drought-induced shortages of raw materials or of the fuelwood needed to cook them. Wars and other social ills may exacerbate scarcities caused by changing climatic conditions.

Shortages of raw materials adversely affect the ability of the small-scale producer to guarantee supplies of the processed product and, hence, to stay in business. For this very reason, local people often display considerable resourcefulness in dealing with scarcity. For example, the shortage of fuelwood in Nigeria has prompted the women who produce *dawadawa* to use soybean rather than the traditional locust bean to make this product. The cooking time for soybeans is shorter than for locust beans (Waters-Bayer, 1989). Women in Ethiopia mix sorghum or millet in when making the sour pancake *injera*, although the preferred ingredient is teff, which is very expensive in times of drought. The besieged Tamil population of northern Sri Lanka have come up with several new processes and products, including *pinnttu*, which is a dried palmyrah fruit juice, ground neem flowers, dried vegetables and smoked meat and fish.

## Quality control

Inadequate quality control may have an adverse effect on local demand for the product. It may also discourage innovation, since producers tend to rely on tried and trusted methods.

In modern industrial applications, the fermentation equipment and processes are controlled using expensive technology, resulting in a consistent product of known quality. Traditional practices, in contrast, take place in a less predictable environment in which the risk of failure is greater. Examples of fermentation which do not produce the desired results include sour beer, ropey bread and the contamination of yoghurt by coliforms (bacteria associated with faeces).

It is often asserted that traditional food products are unhygienic and/or adulterated. This is undoubtedly true in some cases, but evidence from several sources suggests that the problem has been overstated. In Indonesia, a study by the Equity Policy Centre found that only three per cent of street food enterprises were dirty (Tinker, 1988). Studies in Nepal and India (Calcutta) have shown that the microbiological counts in street foods are not a serious problem. Many fermented food products are inherently safe due to their low moisture contents or high acidity. *Achar, gari* and *gundruk* are examples. Several of the steps used in traditional processes are designed to reduce the chances of contamination. These include soaking, cooking and adding salt (Paredes Lopez, 1992).

Taste and texture are further quality control issues. Using fresh raw materials and deploying a high level of culinary skills handed down through the generations, small-scale producers frequently have an advantage over their competitors in the industrial sector in this area. Nevertheless, some rural foods are regarded as unrefined by city dwellers. Traditional beers are an example: *pito*, the traditional beer of Ghana, is much more popular in rural than in urban areas.

## Competition

In some countries and for some commodities, rural food processors now face increasing competition from agro-industry. The changing pattern of food consumption, while opening up new opportunities for small-scale processors, is also reducing the demand for certain traditional fermented products.

In Nepal, for instance, the traditional dish of rice and *gundruk* is being replaced by factory-made packaged noodles (Khanai, 1991). This has reduced the opportunity for rural producers to generate income. As the means of production are concentrated, the knowledge of how to make the traditional fermented product, previously vested in millions of rural women who no longer put it into daily practice, is being gradually lost.

In the future, small-scale processors will have to become still more responsive to changing market conditions. Flexibility is one of the major advantages of being small, so in itself this challenge does not present great problems. But where government policies constrain the natural innovativeness of the small-scale entrepreneur and stifle private enterprise, they will have to be changed.

## Marketing

For all business people, it is essential to find out exactly what the market wants and the price it is willing to pay, then to acquire the means and apply the skills needed to make and sell the product. Food processing is no different. In the past, rural producers of biotechnological food products have gathered this information through word of mouth, talking in the market-place or tea-house, for example. With the rapid escalation in the scale and sophistication of markets in developing countries, these informal channels are no longer sufficient.

Many development organisations have tried to address this problem, not often very successfully. Perhaps the most promising initiatives in this area are those taken by non-governmental organisations (NGOs). In Bangladesh, the Mennonite Central Committee has set up a marketing organisation to promote the produce of small groups of processors. It is presently investigating the potential of *achar* (pickled vegetables). In Pakistan, the Aga Khan Rural Support Organization has helped producers to establish regional marketing associations (Durrani, 1992). The UK-based Intermediate Technology Development Group includes marketing in all the food processing training courses it runs for field workers.

A recent project coordinated by the International Irrigation Management Institute (IIMI) in Sri Lanka has adopted a novel approach, seeking to link producers' groups with banks that can provide loans to start operations, against forward contracts with Colombo-based companies to buy the produce.

## Product status

In a few cases, local products have fallen into decline not for quality reasons but solely because they are regarded as being of lower status than industrial products. The classic instance of this was the creation of a false market for formula milk by the company Nestlé, exposed by *New Internationalist* during the 1970s. The company had successfully indoctrinated professional medical staff, who were advising mothers to use the modern Western product rather than breast-feed, which was regarded as backward.

The "Coca Cola syndrome" plays a role here. Attractive, brightly coloured packaging is one of the big selling points of the newer products. The product looks better and has a more glamorous image.

It is important not to confuse issues of status with those of quality. In some countries there has been a switch to drinking imported beer rather than sorghum or banana beer, but this may reflect a genuine change in consumers' tastes as incomes rise, rather than solely a perception of product status.

## Policy environment

The policy environment in many countries discriminates against the small-scale food processor. Quality standards and legislation favour the larger enterprises. Credit is rarely easy to obtain for the small business. Women, who are the main practitioners of traditional food processing, have seldom been included in research and development projects, although the record in this respect has improved markedly in recent years.

One policy constraint that does not apply to food processing is that of price. In many countries the producers of raw materials are forced to sell a certain proportion of their harvest at an artificially low price to the government, but such restrictions seldom apply to processed products, the market for which is usually free.

# Research and development needs

Research and development needs fall into four broad areas:

- improving understanding of the technologies traditionally used in fermented food processing,
- research to improve product quality and quantity,
- more and better transfer of traditional and improved food processing technology,
- selective policy interventions to promote small-scale production and marketing.

## Improved understanding

The basic mechanisms underlying the biotechnologies used in food processing are well understood. So too are the processes and equipment required to make the world's most widely available fermentation products. Examples are major dairy products, vegetable pickles, wine, beer and spirits. Generally, however, research has been restricted to those products and processes with large-scale commercial applications. With some notable exceptions (such as cassava), less work has been done on the processing of commodities of importance to poor people in developing countries. The case of Sudan is typical (Box 5.3).

More research is therefore needed on the practices and processes used by such people. Such research as has been conducted so far has been largely descriptive in nature, with few quantitative data. Information is needed on the contribution of fermented foods to diets in the developing world, and on their significance in terms of costs, incomes and labour requirements.

Diagnostic research should explore the existing knowledge systems of local people, helping to make these systems more visible and better appreciated. It should also emphasise the identification, with local people, of the problems they face in improving their products and processes. This information is essential for supporting research on possible technological improvements.

## Improved product quality and quantity

Using the diagnostic research outlined above as a basis, participatory research with producers is needed on key components that appear likely to lead to improved product quality and/or quantity. Much of this research will be adaptive in nature, refining technology developed elsewhere for use in local circumstances. Much of it, too, will be conducted with women, the major practitioners of biotechnological food processing virtually throughout the world, and the major potential beneficiaries of such research.

---

**Box 5.3**
***A plea for new priorities***

In Sudan, the traditional biotechnologies of food fermentation have not been a priority on the research agendas of funding agencies. Instead, much attention is given to recombinant DNA techniques and other modern genetic tools. Most of the funds for biotechnology research and development are provided by multinational companies seeking large profits. Why should they be interested in poor people's foods?

To ensure equitable development, research should be geared to the needs of poor people. Strengthening the role of fermented foods and beverages in the struggle against malnutrition, seasonal food shortages and famine should be a major priority. Modern biotechnology could be applied to improve first the raw materials used in these foods, then the fermentation processes themselves. One area of great potential is the development of genetically engineered starter micro-organisms. These might trigger the release of larger quantities of vitamins and amino acids in fermented food, helping to improve its nutritional value.

The benefits would not flow only one way. The micro-organisms used by rural women might enrich many areas of science-based biotechnology. For example, the yeasts used for *duma* (a form of mead) have already proved to be vigorous fermenters of cane molasses.

Source: Dirar (1993).

It will be important to give priority to those commodities and products that offer the greatest income-earning opportunities for poor people and that promise major gains in productivity, quality, efficiency and incomes. Some commodities, in particular, have considerable potential for the development of new products and markets if processing efficiency can be improved. Cassava is the most obvious example, with opportunities in the manufacture of starch, animal feeds, flour for bread-making, paper making, glue and several other applications. Among existing fermented products, *idli* in South India, *gundruk* in Nepal and fish sauces from South-East Asia are examples of products with a wider potential market than that currently exploited. In the case of *idli*, an American company is currently attempting to buy rights to the fermentation process, surely an indication of real potential. Several traditional products could contribute considerably to export earnings. For example, *satcora achar,* a pickle made from local citrus fruits in Bangladesh, has found a growing market in the UK among British people of Bangladeshi origin.

Research on improved equipment should focus on the transfer and adaptation for local use of simple low-cost technology successfully applied elsewhere. Helping people find the right solutions in this area could bring major efficiency gains, relieving the drudgery of women. Improved churns for butter making, which appear widely applicable in Africa, are just one example out of many. In developing such technology care must be taken to emphasise the use of locally available materials and to keep investment and operating costs to the minimum.

Supportive research in such areas as improving fuelwood supplies and the efficiency of cooking stoves is already under way but will continue to be needed, especially as markets expand, raising pressures on the natural resource base that supports food processing.

Quality control systems suitable for use by small-scale producers need to be developed and introduced. Such systems should be developed with the people who must understand and apply them, namely producers themselves. The subjects covered should include the construction (or alteration) of food processing workshops, hygiene and safety procedures, product consistency, and compliance with food standard regulations. Training and information materials are already available on some of these subjects (Fellows and Hidellage, 1993).

## Technology transfer

The food processing technologies used by poor people in developing countries are often highly location-specific. They are used by a particular group of people, using locally available commodities and equipment to cater for local tastes.

Nevertheless, there are opportunities for transferring some technologies to new users. Ugandan women have already learnt new cassava processing techniques through exchange visits to Ghana (Simwogevere, 1993). There is an opportunity to transfer the uses of tamarind to Zimbabwe, where the fruit is available but under-utilised at present, from several other countries in Africa, Asia or Latin America (Mpande and Mpofu, 1993).

Successful technology transfer is based on a firm understanding of how people interpret and assimilate new technical inform-ation. This is especially important in the early stages of research and development, when potential users are not yet fully commit-ted. At a later stage, learning by doing should ensure strong loyalty to the new activity, especially if its benefits are felt quickly.

Despite the location-specific nature of most food processing technologies, some have spread well beyond their area of origin. In the case of fermentation techniques in Sudan, historical evi-dence shows that as women travelled they shared their skills across the continent all the way to West Africa (Dirar, 1993). It may be worth devoting some research effort to understanding the principles that underlie the spontaneous spread of technology.

## Policy interventions

How can governments best support and enhance small-scale food processing and marketing?

Policies that encourage small businesses in the private sector are a good starting point. Private enterprise is still stifled by a mix of poor exchange rate policies, high taxation, unnecessary red tape and a corrupt ruling class in a number of African countries. Small-scale producers, and women in particular, badly need bet-ter access to credit.

The status problem has important policy implications. In the few cases where powerful advertising campaigns or other tech-niques of persuasion by large companies have created a false market for new products that are actually inferior to traditional ones, governments may need to take strong corrective action. Where such products are imported, they can be banned. Professionals in the food industry and in development circles generally could be targeted by government-sponsored promo-tional campaigns to right the balance, stressing the advantages of the displaced local product.

However, in cases where the decline of a traditional product reflects a genuine change in tastes it would be fruitless to attempt to tackle the status issue directly. The best returns to research and development will be achieved by improving prod-uct quality. This will enable producers to sell more of the prod-uct at higher prices and its status will improve as a result. In such cases marketing efforts could be supported through sub-sidised advertising campaigns on national radio and television for groups of small-scale entrepreneurs making improved ver-sions of traditional products.

# References

Adeyemi, I.A. 1993, 'Making the most of Nigerian *ogi*.', *Food Chain* 8: 5. Intermediate Technology, Rugby, UK

Aguhob, S. and Axtell, B. 1994, *Dairy Processing: Food Cycle Technology Source Book,*. UNIFEM, New York

Ashenafi, M. 1992, 'Microbiology of Ethiopian *ayib*', in Gaden, E.L. (ed), *Applications of Biotechnology to Traditional Fermented Foods,* National Academy Press/Board on Science and Technology for International Development (BOSTID), Washington DC, USA

Beddows, C.G. 1985, 'Fermented fish and fish products', in Wood, B.J. (ed), *Microbiology of Fermented Foods* Vol 2. Elsevier, Amsterdam

CFTRI. 1986, 'Traditional foods: Some products and technologies', Central Food Technological Research Institute, Mysore, India

Davies, G. 1994, 'Domestic banana beer production in the Mpigi District, Uganda', paper submitted to the Contest on Rural People's Biotechnology, ETC Foundation, Leusden, Netherlands

Dirar, H. 1992, 'Sudan's fermented food heritage', in Gaden, E.L. (ed), *Applications of Biotechnology to Traditional Fermented Foods,* National Academy Press/Board on Science and Technology for International Development (BOSTID), Washington DC, USA

Dirar, H. 1993, 'Fermented foods in Sudan', in *Appropriate Technology* 20 (2): 24

Durrani, S. 1992, 'Apricot magic in Pakistan', *Food Chain* (7): 10, Intermediate Technology, Rugby, UK

Ezedinma, C.I. and Igbinnosa, I. 1993, 'The production of *ugba*', paper submitted to the Contest on Rural People's Biotechnology, ETC Foundation, Leusden, Netherlands

Fellows, P.F. and Hidellage, V. 1993, *Making Safe Food,* IT Publications, London

Gaden, E.L. (ed), 1992, *Applications of Biotechnology to Traditional Fermented Foods,* National Academy Press/Board on Science and Technology for International Development (BOSTID), Washington DC, USA

Harper, D.B. and Collins, M.A. 1992, 'Leaf and seed fermentations of Sudan', in Gaden, E.L. (ed), *Applications of Biotechnology to Traditional Fermented Foods,* National Academy Press/Board on Science and Technology for International Development (BOSTID), Washington DC, USA

Iji, P. 1993, 'Fermentation and distillation of local gin', paper submitted to the Contest on Rural People's Biotechnology, ETC Foundation, Leusden, Netherlands

Karki, T. 1986, 'Some Nepalese fermented foods and beverages', in *Traditional Foods: Some Products and Technologies,* CFTRI, Mysore, India

Khanai, P. 1991, 'Hello chou chou! Goodbye dal bhat!' in *Food Chain* 2: 7, Intermediate Technology, Rugby, UK

Koleoso, O.A. and Kuboye, A.O. 1986, 'Traditional food and beverage technology of Nigeria', in *Traditional Foods: Some Products and Technologies,* CFTRI, Mysore, India

Kumar, S. 1993, 'Indian fermented milk', paper submitted to the Contest on Rural People's Biotechnology, ETC Foundation, Leusden, Netherlands

Lücke, F.K. 1985, 'Fermented sausages', in Wood, B.J. (ed), *Microbiology of Fermented Foods* Vol 2. Elsevier, Amsterdam

Martin, A. 1993, 'Cassava fermentation into *fufu*', paper submitted to the Contest on Rural People's Biotechnology, ETC Foundation, Leusden, Netherlands

Menea, K.B. and Bishosha, A.K. 1993, 'Transformation of cassava leaves by Lokele people in the Republic of Zaire', paper submitted to the Contest on Rural People's Biotechnology, ETC Foundation, Leusden, Netherlands

Mpande, R. and Mpofu, N. 1993, 'Coping strategies', in: *Appropriate Technology* 20 (2): 22

Munyanganizi, B. 1975, 'La technologie de l'extraction du jus de bananes et sa vinification', doctoral dissertation, State University of Gembloux, Belgium

Mwangi , D.K. and Mwangi, A.N. 1993, 'Improved food fermentation using amaranth', paper submitted to the Contest on Rural People's Biotechnology, ETC Foundation, Leusden, Netherlands

Oberman, H. 1985, 'Fermented milks', in Wood, B. (ed), *Microbiology of Fermented Foods* Vol 1. Elsevier, Amsterdam

Odunfa, S.A. 1985, 'African fermented foods', in Wood, B. (ed), *Microbiology of Fermented Foods* Vol 2. Elsevier, Amsterdam

Paredes Lopez, O. 1992, 'Nutrition and safety considerations', in Gaden, E.L. (ed), *Applications of Biotechnology to Traditional Foods,* National Academy Press/Board on Science and Technology for International Development (BOSTID), Washington DC, USA

Sanchez, P.C. 1986, 'Traditional fermented foods of the Philippines', in *Traditional Foods: Some Products and Technologies,* CFTRI, Mysore, India

Simwogevere, E. 1993, 'Cassava use and processing', in *Appropriate Technology* 20 (2): 23

Sokari, T.G. 1992, 'Improving the nutritional quality of *ogi* and *gari.*', in Gaden, E.L. (ed), *Applications of Biotechnology to Traditional Fermented Foods,* National Academy Press/Board on Science and Technology for International Development (BOSTID), Washington DC, USA

Steinkraus, K.H. 1985, 'Bio-enrichment: Production of vitamins in fermented foods', in Wood, B. (ed), *Microbiology of Fermented Foods* Vol 1. Elsevier, Amsterdam

Steinkraus, K.H. 1992, 'Lactic acid fermentations', in *Applications of Biotechnology to Traditional Fermented Foods,* National Academy Press/Board on Science and Technology for International Development (BOSTID), Washington DC, USA

Svanberg, U. 1992, 'Fermentation of cereals: Traditional household technology with nutritional potential for young children', IDRC Currents 2: 21–24. IDRC Canada

Thro, A-M. 1993, 'Cassava sour starch', paper submitted to the Contest on Rural People's Biotechnology, ETC Foundation, Leusden, Netherlands

Tinker, I. 1988, 'The case for legalising street foods',*Courier* 10: 72–74

Tinker, I. and Cohen. M, '1986. Street foods as income for the poor', Equity Policy Centre, Washington DC, USA

Vaughn, R.H. 1985, 'Microbiology of vegetable fermentations', in Wood, B. (ed), *Microbiology of Fermented Foods* Vol 1. Elsevier, Amsterdam

Waters-Bayer, A. 1989, 'Soybean *dawadawa*: An innovation by Nigerian women', ILEIA Newsletter No 3: 17

Winaro, F.G. 1986, 'Traditional technologies of Indonesia, with special attention to fermented foods', in *Traditional Foods: Some Products and Technologies,* CFTRI, Mysore, India

Wood, B.J. (ed), 1985, *Microbiology of Fermented Foods* Vols 1 and 2, Elsevier, Amsterdam

Wood, B.J. and Hodge, M.M. 1985, 'Yeast-lactic acid bacteria interactions', in Wood, B.J. (ed), *Microbiology of Fermented Foods* Vol 1. Elsevier, Amsterdam

# 6 Crop Genetic Resources

*Walter S. de Boef, Trygve Berg and Bertus Haverkort*

## Introduction

The biotechnology developed and used by rural people includes the selection and breeding of plants and animals. This technology has led to the domestication, spread and adaptation of thousands of species in diverse environments all over the world. It has also entailed the accumulation of an enormous wealth of genetic variability, much of which is still being maintained and managed by local farmer-breeders. Considering the difference between wild and cultivated forms of the same species, this achievement goes far beyond many other human accomplishments. The genetic resources of domesticated plants and animals are perhaps the most valuable asset ever generated through human activity.

In modern societies, most plant breeding has been taken over by professionals, many of them working for private companies. Some plant breeders have, in the past, explicitly or implicitly dismissed farmers' efforts at plant breeding as backward or primitive. In recent years, however, a growing number of researchers have challenged such prejudices. Investigation of the practices associated with farmers' plant breeding has revealed a rich and dynamic culture that is still a potent force in local agricultural development.

The Keystone Dialogue Series on Plant Genetic Resources was one of the first international fora to recognise the informal crop development system as a counterpart to the formal, institutional system. Following Keystone Center (1991), Berg et al. (1991) and Hardon and de Boef (1993) described these two independent yet complementary systems of crop development as follows:

- In the informal or local system, the farmer acquires seeds by saving them on his or her own farm or from other farmers who have done so. Seed saving involves selection, both natural and conscious, and results in what may be called the dynamic conservation of landraces, since conservation here entails continuing evolution. The informal system relies on, and indeed embodies, the skills of farmers in maintaining, enriching and utilising crop diversity. The main selection criteria used are yield and yield stability, risk avoidance, low dependence on external inputs, and a range of quality factors associated with storage, cooking characteristics and taste.

● The formal system has two distinct components. The profit-oriented private sector concentrates on yield-increasing technology, often coupled with the use of agro-chemicals. It caters mainly for the needs of larger farmers living in the higher-potential (usually irrigated) areas, who can afford such inputs. The public sector, consisting of international and national research systems and a few regional bodies, also produces varieties for use in the high-potential areas, but also caters, if less successfully, for the needs of resource-poor farmers living in more marginal rainfed areas, where the conditions for production are less predictable. In this sector yield increases are still an important breeding objective, but there is greater emphasis on yield stability and on a range of traits that allow the use of external inputs to be avoided or reduced, such as resistance/tolerance to pests and diseases, drought and toxic soil conditions. In both sectors, seeds are multiplied by the seed industry and distributed in what has been widely criticised in the past as a linear model of transfer, with insufficient feedback from users. The private sector predominates in the industrial countries of the North. However, in the South, governmental institutions and the international centres of the Consultative Group on International Agricultural Research (CGIAR) are the main actors.

The informal system is still dominant in many areas, especially those with marginal environments for crop production. The main reason for this is that few modern improved varieties have yet been developed for such environments. In most other areas the formal and informal systems co-exist and interact. Where modern varieties are adopted, they are usually integrated into the local seed production system.

## The informal system

Man has been a hunter and gatherer almost throughout the 2.5 million years of his earthly existence. It was only 10 000 years ago that the domestication of plants and animals began. This was the start of farmers' efforts at selection, based on the distinction between more or less useful individuals and species. Initially, extensive production systems such as shifting cultivation and pastoralism predominated. Later, diverse forms of sedentary, integrated and more intensive farming systems developed. Specialised farms with monocropping, using a high level of external inputs, have become widespread in temperate areas only in this century. In large parts of the tropics and subtropics, they remain the exception rather than the rule.

Hardon and de Boef (1993) use the term local crop development to describe the continuous and dynamic interaction between farmer and crop that is the central characteristic of the informal system. The main components of local crop development are the maintenance of local varieties (conservation through utilisation), their enhancement (through selection and enrichment with exotic materials) and the seed system (production, selection, treatment, storage and exchange) (Amanor et al., 1993). In most traditional societies, local crop development is practised by both men and women, but women often have the main responsibility for the selection and safe keeping of seeds.

Another key feature of local crop development is its maintenance of diversity, both between and within crop species. This diversity is the subject of a wealth of knowledge accumulated by its custodians. Pereira (1990) cites the example of a 12-year-old girl belonging to the Warli tribe of India, who knew the names and uses of over 100 herbs, shrubs and trees used for fibre, fuel, lighting, food and medicinal purposes. Not infrequently the knowledge of farmers is equal or even superior to that of researchers. A more detailed account of indigenous knowledge in relation to the informal crop development system is given in de Boef et al. (1993).

## Diversity among crops: minor species

The deliberate mixing of a wide range of crops in combination with different animal species has several advantages when appropriate combinations, complementary in their functions and resource requirements, are devised. Mixtures can lead to highly efficient systems with relatively stable yields. Minor crops and semi-domesticated plant species play an important role alongside the major food crops.

Shifting cultivation systems, as well as more permanent mixed smallholder systems, are renowned for their deployment of large numbers of plant species. Farmers use not only the area cultivated (which may vary from one year to the next) but also the surrounding fallow areas, bush and forest, all of which serve different purposes. In recent history, the physically central place in the farming system of a few major crops has been paralleled by a corresponding emphasis in research and development, often to the near exclusion of other commodities seen as peripheral. Prescott-Allen and Prescott-Allen (1990) consulted various sources to arrive at an estimate of the number of species accounting for most of the world's plant food supply. They found that five sources listed only seven to 26 species, but that by using statistics of the Food and Agriculture Organisation (FAO) on per capita food supply in 146 countries and counting those species required to make up 90% of each country's plant food supply, they reached a list of 103 species. If their study had been conducted at local community level, the numbers would have been higher still. Many communities depend on species which do not appear at all in national statistics or, if they do appear, are lumped together as "minor crops". Thus modern science and technology have put virtually all their eggs in one basket, that of the 20 or so species deemed to account for the bulk of current world food supplies. The remaining multitude of crops, and the people who use them, depend on indigenous biotechnology for their survival and continuing improvement.

Many examples of minor crops could be cited. Among the Lua of Northern Thailand, about 120 crops are grown; 75 food crops, 21 medicinal crops, 20 plants for ceremonial purposes and seven for weaving and/or dyeing. The fallow swiddens are also highly productive, with over 300 species either used for grazing or collected for a range of purposes (Kunstadter, cited in McNeely, 1989). Seetharam et al. (1990) includes a review of the under-researched small millets and several papers documenting their local importance and potential. Van der Maesen et al. (1989) document 21 major pulses, three minor ones and 69 other species occasionally used as pulses but mainly exploited for other purposes (medicine, tannin, dyes, timber or forage). This study also found 395 under-researched species of fruits and nuts.

The bias of research towards a few favoured species has its roots in the history of colonialism (Box 6.1), and is perpetuated by the continuing shortage of resources available to national and international research systems. Vegetables provide a classic example; most research efforts are devoted to "modern", temperate, vegetables from the North, to the exclusion of indigenous vegetables.

Minor crops have great potential to contribute to agricultural development, especially for farmers in marginal areas. NGOs active in the Andes have put considerable energy into projects to maintain and strengthen the cultivation of minor Andean crops. Often, these projects have been linked to political initiatives to strengthen the position of indigenous people (Tapia and Rosas, 1993; CLADES et al., 1994). This type of project can help to draw these crops to the attention of researchers in the formal system, bringing new resources to bear on their improvement.

---

**Box 6.1**
**The impact of botanical colonialism on Andean crops**

At the time of the Spanish conquest of South America in the sixteenth century, the Inca had domesticated as many as 70 crop species, including a wealth of roots, grains, legumes, vegetables, fruits and nuts. The conquerors considered the natives to be backward and uncreative. Crown and Church prized silver and souls, not plants. Crops that had held honoured positions in Indian society for thousands of years were replaced by European species.

But in the higher Andean areas less affected by the conquerors, indigenous crops such as *oca, maca, tarwi, nunjas* and *lucuma* continued to be grown for the next 500 years. The Andean population has maintained these crops in the face of neglect and even scorn by much of the society around them. A study by the National Research Council (1989) of the USA described some 30 Andean species which deserve more widespread use and appreciation beyond the small area to which they are now restricted. These crops gave acceptable yields, demonstrated a remarkable adaptability to different environments and often had high nutritional value.

Source: National Research Council (1989).

## Diversity within crops: landraces

Traditional farming systems often exhibit considerable genetic diversity within crop species, especially the landraces developed by farmers over the generations within the system. McNeely (1989) reports a Nepalese farmer who grew 20 varieties of rice. Although illiterate, the farmer knew exactly which variety he was going to plant where, and why. He had a detailed understanding of how such variables as slope, pests, rainfall and temperature affected the yields of each.

Maintained in a dynamic process involving both natural and human selection, landraces tend to be well adapted to the environment in which they have evolved. Many landraces can still be found in marginal areas. When grown under the stressful conditions that typify such areas, they often appear superior or at least equal to modern varieties in terms of yield (Weltzien and Fischbeck, 1990). The variability inherent in landraces serves the same purpose as the crop mixtures grown by farmers; to escape total crop failure by reducing vulnerability to specific diseases, pests and climatic stresses (Box 6.2).

Most landraces have been subject to selection by farmers under variable seasonal conditions within the same environment. This kind of selection enhances yield stability within the environment despite fluctuations in rainfall and temperature (Ceccarelli et al., 1992).

Yield stability in landraces can be achieved in two ways. A given genotype may have a low genotype × environment interaction, enabling it to yield under both stressful and optimum conditions. Alternatively, in a genetically mixed population, genotype substitution may occur, such that plants which fail to produce under one set of conditions yield well under different conditions (Hodgkin et al., 1993). In self-pollinating crops such as rice and common bean, genotype mixtures tend to consist of several pure lines, with a very low frequency of cross-pollinations occurring. In open-pollinated species such as maize, or in crops with a high level of cross-pollination such as sorghum, landrace populations are subject to extensive intra- and inter-varietal recombinations (Brown, 1978; 1979).

## The seed system

The seed system is the central component of local crop development, through which its other functions (maintenance, enhancement) are mediated. It consists of the selection of the reproductive materials, seeds, tubers, and so on, followed by their treatment and storage until planting. In some traditional societies, specialised farmer-breeders may operate and seeds are more widely exchanged and distributed through extensive regional networks. Our discussion here focuses on the tasks of selection and dissemination, which are the central ones from the point of view of maintaining biodiversity.

---

*Box 6.2*
*Potato landrace diversity in the Andes*

Farmers in the Andean *altiplano* commonly plant up to 20 different potato varieties in the same plot. They know each variety by name and where it comes from. They know which varieties are susceptible to which diseases and climatic stresses, and can describe each in terms of the characteristics of its flowers and leaves and the taste of its tubers.

A farmer explained: "Maybe some of the varieties won't yield so much in this dry year, but we still have others which can put up with some dryness. In a wet year, it can be just the opposite, and we are glad of the potatoes that are not so liable to rot. Some varieties are more resistant to frost, others to cutworms".

Source: Benzing (1989).

## Seed selection

Almekinders et al. (1994) name four sources from which farmers may obtain their seeds: their own harvested material, other farmers, the local grain market and the formal sector. Farmers' own seed is the cheapest. It is also of known quality and, unless in times of disaster, readily available. Farmers turn to other sources only when they have been unable to save their own seed or when their seed has degenerated. For instance, Sperling et al. (1993) estimated that, even before the outbreak of civil war in Rwanda, nearly half of poor farmers in the south of the country bought over 90% of their bean seed on the local grain market during the main growing season. They did so because they had no choice but to consume their own seed.

The levels of skill displayed by farming communities in selecting their seed often rival those in the formal system. In a comparison between farmers' seeds of the common bean (*Phaseolus* spp) and those produced by researchers, Janssen et al. (1992) found that farmers' seeds produced similar or higher yields in 11 out of 13 cases.

Seed selection on the farm leads, as we have seen, to the continuing adaptation of local cultivars. Farmers deliberately maintain a degree of variation as a buffer against constraints such as climatic fluctuations, changing populations of pests and the occurrence of diseases. Farmers' seed selection is also integrated with the social, cultural and religious life of the community. Box 6.3 provides an example.

Seed selection is normally preceded by extensive discussions, both within the farm family and with neighbours. Any family

---

*Box 6.3*
**Selection of rice by the Mende people of Sierra Leone**

The selection of rice seeds by the Mende farmers of Sierra Leone has resulted in a systematic grouping of the main rice types. Early-maturing types are picked as they ripen by farmers anxious to provide their families with a little extra food during the hungry season that precedes the main harvest. Longer-duration types are left in the field for the gleaners after harvest. In this way, over a long period of time, Mende rice germplasm has separated into three distinct maturity classes. There is a definite prejudice against consuming the long-duration varieties. Harvested last, these varieties are not considered vital to subsistence and have a low status in the village "moral economy". They are the rices that people can afford to sell because they are not needed to build up the community.

By contrast, acquiring short-duration seed types and sharing them with family and friends is a significant social act, because it repairs the social fabric damaged by climatic fluctuations and other misfortunes. Farmers search for short-duration types because these represent a practical answer to the social evil of indebtedness caused by pre-harvest hunger. Suitable planting materials are seized upon, segregated and regularly tested in small trial plots. Farmers are explicit about the need to maintain the diversity of rice germplasm through such experimentation and exchange. They say: "It is the nature of rice and circumstances to change". No farmer believes in an ideal variety to which he or she will remain committed for life. The generations of rice are like the generations of humans: no child is an exact copy of either parent. Rice changes within the framework governed by ancestral forces, the guardians of the moral order.

Source: Longley and Richards (1993).

member may make observations of crop performance, looking at the crop during weeding or other activities and noting any interesting variations. Children posted in the field as bird scarers may come home with ideas on selection. Similarly, a good crop stand is often noticed by neighbours and becomes a subject of conversation within the community. Observations and ideas, as well as the seeds themselves, are extensively exchanged.

Ideas on selection are raised for consideration by the person who must actually take the final decisions, accepting responsibility for the next generation of plants. Fittingly, this is often the woman of the household. Women have the knowledge essential for seed selection because of their many other crop-related roles and tasks. They may be responsible for various crop husbandry tasks as well as for harvesting, and usually conduct all post-harvest operations such as processing, cooking, storage and use of crop residues. Post-harvest properties are especially important in subsistence production systems. Women have the power to refuse any variety found wanting in this respect.

Market-oriented production often leads to the adoption of improved seeds bred by the formal system, with the result that seed selection becomes a less significant activity. Women's assessments may become less important, since criteria such as yield and price rather than taste and cooking quality become the overriding concerns.

However, even when the formal system takes over, traditional seed selection may still survive, albeit in a diminished form. Women often insist on keeping some of the older varieties for home consumption because of their culinary qualities. In the home garden, which is the women's domain, several old species may be inconspicuously tended and nurtured. In keeping these old species alive, these women keep traditional knowledge and culture, the old biotechnology, alive.

## Specialised seed production and distribution

Within communities, specialised seed producers may sometimes operate, supplying farmers with high-quality seeds and/or seeds of new varieties (Osman, 1990). These farmers may be a source of expertise on the adaptability of different varieties to local conditions. They grow seed in large numbers of dedicated plots.

Traditional seed-producing areas can be a source of healthy and vigorous seed to replace degenerated stocks. In the Andes, seed potatoes are mostly produced at higher altitudes with lower temperatures, conditions in which the incidence of virus infections is reduced (Prain and Scheidegger, 1988; Rhoades, 1985). Villages at these altitudes may specialise in seed production, numbering several highly skilled farmer-producers among their inhabitants (Linnemann and Siemonsma, 1989).

Such villages may be sources of new varieties, which are subsequently spread through networks based on family relations and other social ties. Migrant labour channels are often a source of new seeds. In southern Africa, men working in the mines of Botswana and South Africa often bring home foreign seeds and new crop varieties when they return to their own countries (Van Oosterhout, personal communication).

Farmers in certain areas of Ethiopia's Tigray and Gondor provinces are renowned for maintaining and producing "elite" material of local landraces. In this country local seed systems have evolved into extensive regional networks. Seeds are widely exchanged at markets, where an assortment of varieties adapted to different environmental conditions is available (Worede and Mekbib, 1993; Mekbib et al., 1993).

## Low-external-input agriculture and farmer experimentation

In large parts of the tropics, the physical environment, commercial infrastructure and/or price ratios between external inputs and farm outputs do not allow the use of large quantities of purchased inputs, especially agro-chemicals. Wolf (1986) estimates that some 1.4 billion people, or about a quarter of the world's population, depend for their livelihoods on low-external-input agriculture (LEIA). The area under LEIA is expanding as the rural population increases and external inputs become more expensive.

In some areas of the Philippines, farmers are now turning to practices with low levels of external inputs after several decades of farming with fertilisers and pesticides. An economist at the Manila-based International Rice Research Institute (IRRI), the origin of much of the modern technology, comments: "Given low prices, declining or stagnant yields and increasing input costs, the net income and welfare of the rice farmer have been declining. The prospects for improvement in this situation are not bright." (Pingali, 1993).

Farmers switching to LEIA use locally available seeds, including selections from local breeding programmes and commercial varieties, as well as traditional land races. The result is a revival of local crop development and farmer experimentation (see Box 6.4).

Coffman and Smith (1991) consider a process in which cultivars are adapted to fit the environment instead of the environment being altered to fit the cultivars as crucial for the development of sustainable agriculture. The current renewal of interest in LEIA, with its shift away from the use of chemicals, represents an important contribution to this aim.

Both LEIA and farmer experimentation tend to flourish in the absence of a formal research system. A further factor affecting experimentation is the migration of farmers to new areas. Perhaps the most striking example of this is the establishment of settler agriculture in North America during the last century. The seeds the immigrants brought with them were often not well adapted to environments in the New World. The need for experimentation

**Box 6.4**
***The revival of farmer
experimentation in the
Philippines***

Farmers adopting Green
Revolution technology in the
Philippines were taught that
modern varieties were pure.
Extension services advised
roguing off-types before harvest,
to prevent them from producing
seeds. For more than 20 years,
nearly all farmers followed this
recommendation.

The old idea of off-types as a
resource rather than a nuisance
resurfaced on the island of
Mindanao in 1985. That year,
one farmer, Eulogio "Gipo" Sase
Jr, planted seeds of an IRRI variety
which he had got from another
farmer. He discovered an off-type
with a dark green colour and
marked it with a stick. His water
buffalo ate part of it, but he
managed to save four panicles. A
month later he planted the seeds

in a small plot after soaking them
in water for three days. He
obtained an average of 42 tillers
per plant and a harvest of 30 kg
of seed. This he planted again,
this time harvesting 127 bags.
Then he started disseminating the
new seed under the name
Bordagol.

Bordagol is now a popular new
rice variety. It is reputed to yield
more than the original IRRI variety
and to have a better taste. In the
beginning it was also resistant
to the Tungro virus, but
resistance broke down in 1989.
According to Gipo, further
off-types started to appear in
the second or third generation of
the new seeds. Many farmers
now grow selections from within
Bordagol.

The case of Bordagol may have
prompted a change in farmers'
attitudes towards off-types and
variations within a variety. When
the Community-based Native
Seed Research Center in

Cotabato, also on Mindanao,
recently began distributing seeds
of traditional varieties, they found
that farmers were very interested
in seed selection. "It's as exciting
as panning for gold", one of
them said (he had previously
been a gold panner). Several
farmers already have their own
selections, and are demanding
more varieties for trials. Generally,
farmers prefer heterogeneous
varieties because these provide
more opportunities for selection.
They frequently discuss the merits
and shortcomings of different
varieties and some have started
making crosses to combine
desirable traits from different
sources (Berg, 1995)

The notion, held by some, of
farmers' breeding as something
gone for ever once modern
varieties are adopted, has proved
false.

Source: Berg and Alcid (1995).

and selection was recognised by the US government, which instructed its diplomatic missions all over the world to collect the seeds of potentially useful crops. Since the government at that time had no experiment stations, all the seeds had to be distributed directly to farmers, leaving them wholly responsible for testing and selection. These activities assumed enormous proportions. The number of packages mailed out to farmers rose from around 300 000 in 1862 to over 20 million in 1897 (Fowler, 1994). This stimulated a build-up in farmers' research probably unparalleled elsewhere in modern times. This research went swiftly into decline when the public sector and the private seed industry took over from farmers early this century.

## The formal system

The formal crop development system also has distinctive components, including the collection and conservation of genetic resources, their improvement through selection and breeding, and their multiplication and dissemination. In contrast to the informal sector, these functions are normally separate from one another in the formal sector, each being carried out by a different set of

Landrace collections being dried after the harvest. When the Community-based Native Seed Research Center (CONSERVE) in Cotabato, Philippines began distributing seeds of traditional rice varieties a few years ago, they found farmers very interested in seed selection.
(*Trygve Berg*).

specialists. Our discussion will focus mainly on conservation and improvement, but we will also touch briefly on dissemination.

## Germplasm conservation

The need to conserve crop genetic resources has long been emphasised by the international community. Two complementary approaches have been developed: *ex situ* conservation and *in situ* conservation.

### *Ex situ* conservation

*Ex situ* conservation is effected through genebanks, which store samples of seeds or other planting materials under controlled conditions of temperature and humidity. The aim is to conserve as much as possible of existing genetic diversity, ensuring its availability for future generations. Materials are collected through plant exploration and are briefly described ("passport data") before being stored.

Over the past few decades genebanks have proved vulnerable to several problems, including failing infrastructure (electricity cuts), underfunding and political instability in host countries. In many genebanks the germination rate of accessions now falls well below the generally accepted level of 85%.

An important characteristic of genebanks is that they "freeze" evolution. Because genotypes are taken from their natural environment and are no longer subject to regular germination or regeneration, their continuing adaptation to changing environmental conditions is suspended. If properly stored, genebank accessions can be reproduced with little change after a long period of conservation. Yet if the same population had been allowed to survive *in situ*, it might have undergone considerable evolution. Berg (1995) argues that *in situ* conservation represents a better option since it allows germplasm to co-evolve with diseases and pests, changing farming systems and climatic conditions.

Resistance to diseases is often found in germplasm from areas where there is genetic diversity of both the crop species and the pathogen concerned, leading to co-evolution. Allard (1990) demonstrated that natural selection in a heterogeneous barley population could bring about a high level of durable resistance to the fungal disease barley scald. He concluded that sorting and selecting among the enormous number of possible combinations of available resistance genes would be more efficiently done by natural selection than by geneticists. Barley populations continuously cultivated in the presence of the pathogen had a higher level of resistance than any sample from the genebank. Material in the genebank did possess resistance, but the work of tracing resistance genes, recombining and accumulating them to reach a similar level of resistance to that of the natural population appeared extremely laborious and costly compared to simply screening and utilising the co-evolved natural barley population.

The information on accessions in genebanks tends to be restricted. Passport data rarely include characteristics described by farmers. Plant explorers often spend only a few minutes on each sample they collect. There is no time to chat with farmers and record their knowledge. In contrast, the local crop development system leaves the connection between germplasm and local knowledge unbroken.

Genebanks have also been criticised in the global debate on property rights in relation to genetic materials. For local communities, *ex situ* germplasm collections are effectively extinct. Materials in the bank are made available to plant breeders in the public or private sector, but not to the farmers and communities whence they came.

## *In situ* conservation

*In situ* conservation involves leaving species in their natural habitat, allowing adaptation and evolution to continue. This approach has been adapted from the methods used in natural resource management to conserve semi-wild species or the wild relatives of crop species. In theory at least, it is particularly appropriate for habitats that are under threat and for areas that are still farmed traditionally, where crops are often enriched by gene exchange with wild relatives and weeds.

The problem with *in situ* conservation as conceived at present is that, unless crops are permanently guarded, security is low.

Natural habitats may dry out or disappear under the plough, land-races may be replaced by other varieties, and foreign genetic material may be introduced as an integral part of farmers' strategy to improve local germplasm. Paying farmers cash to conserve *in situ* appears unsustainable financially. Currently, some donor agencies are prepared to foot the bill over short periods, but who will take over once they withdraw? Rigorously applied, the large-scale *in situ* conservation of landraces would, in any case, entail a return to "backward" agricultural systems, which to many scientists and even to many farmers and conservationists, would seem an unacceptable and unworkable proposition (Hardon and de Boef, 1993).

As the Crucible Group (1994) pointed out, *in situ* conservation stands the best chance of working when it is firmly rooted in local communities and their organisations. This approach, which represents an attempt to integrate the conservation efforts of the formal and informal sectors, will be discussed later.

## Germplasm improvement

Approaches to crop improvement in the formal public sector have changed since the early days of the Green Revolution. This section outlines the major shifts in thinking, together with some of the remaining issues.

### Emphasis on resource-poor farmers

The international agricultural research centres have recognised the needs of resource-poor farmers living in rainfed and marginal areas. Breeding objectives at many centres now target the traits of interest to such farmers, including yield stability, resistance to pests/diseases, drought tolerance and low dependence on external inputs. This approach has been developed as a parallel strategy to the continuing generation of yield-maximising technology for high-potential areas. According to Hardon (1995), it requires the development of new breeding methods involving more on-farm testing and greater farmer participation in technology design.

Breeding programmes in national research systems, in contrast, have been somewhat slower to evolve, and still tend to give priority to yield-increasing technologies for the high-potential areas. This is because national systems are often under pressure from policy makers to conduct near-market research with the prospect of a rapid pay-off in line with national development objectives. The difference in perspective between national and international programmes has sometimes created tension between them. NGOs are not the only source of pressure on the international centres, and they sometimes pull in the opposite direction to that of national systems.

Biotic and abiotic stresses have proved difficult to combat using formal plant breeding techniques. Nevertheless, while some gaps remain there has been considerable progress in understanding the physiological mechanisms underlying adaptation to stresses, and in identifying useful sources of resistance or tolerance in farmers' landraces. According to the international centres, this knowledge has formed the basis for the successful development of new crop varieties for several commodities widely grown by resource-poor farmers, including cassava, maize, sorghum, pearl millet, chick-pea, cowpea and pigeonpea. Slow at first, rates of adoption of new technology have apparently begun to quicken in recent years. However, in many cases the centres have yet to produce hard evidence in support of these claims, and it remains unclear whether the very poorest farmers have been able to adopt. A new impact assessment unit established by the CGIAR should help to throw more light on these matters.

A great deal of experimental work remains to be done on identifying promising sources of resistance, especially in wild relatives but also in landraces, and on the use of native populations exhibiting high genetic variability.

## On-station versus on-farm research

The development of new technology in the formal sector relies heavily on selection and testing on the research station, where environmental conditions can be unrealistically favourable compared with those faced by small-scale resource-poor farmers. The terrain is usually flatter, and soils are often inherently more fertile. Water supplies, typically in the form of irrigation, may also be more plentiful and regular.

The development of Green Revolution-type technology was associated with the use of a high level of inputs. In recent years there has been a shift towards lower levels, more in line with the conditions faced by resource-poor farmers. However, research directors sometimes find it hard to monitor all the activities of individual scientists, with the result that the covert overuse of agro-chemicals may persist amongst those most resistant to change.

In recent years there has been a pronounced shift in the formal system towards the testing of new materials on farm at an early stage of the research process. This has come in recognition of the value of obtaining farmers' inputs into the technology design process. Many programmes now invite farmers to the research station to view the genotypes being grown there, play an active part in selection and give the reasons for their choices. This can throw light on farmers' priorities, which can then be fed back into the breeding programme.

One danger of this approach is that, unless a sufficiently broad cross-section of farmers is consulted, it may skew the breeding programme in the direction of specific adaptation to a set of local conditions or preferences not common elsewhere (see below).

### Broad versus specific adaptation

Much of the debate within the formal crop breeding system during the 1980s focused on the environment in which crops should be selected.

Some of the early international breeding programmes associated with the Green Revolution adopted what they called broad adaptation as a major breeding objective. These programmes claimed to develop varieties that would yield well over a wide spectrum of environments. However, most of the researchers they employed were experienced in breeding for areas of high to medium potential. Relatively few understood the more challenging and less predictable conditions associated with marginal areas. As a result, the spectrum of environments to which the new varieties were suited lay too high up the scale and was in fact relatively narrow. It excluded the truly marginal areas, where annual rainfall can be as low as 200 mm.

Under these circumstances, "broad adaptation" turned out to be a misnomer. As Ceccarelli (1989) and others showed, the assumption of these programmes, that varieties with a high yield potential under optimal (on-station, high-input) conditions would also express their superiority under limiting conditions, was not valid for truly marginal environments, where many such varieties failed.

Researchers at the second-generation international research centres still pursued broad adaptation, but did so in a way more in keeping with the needs of marginal areas. They recognised the high inter-annual variability of rainfall in such areas and the multiple biotic stresses to which they were subject.

In most breeding programmes today, broad adaptation remains the second most important selection criterion after yield. For reasons of cost-effectiveness, varieties have to be developed to produce well over a wide range of environments. Although well suited to coping with highly variable stresses such as drought, this approach may eliminate materials that only perform well under specific conditions or that have locally sought-after quality characteristics. These discarded materials may harbour valuable genes that are lost to the plant breeding effort. Where a specific stress is a permanent fixture of the local farming system, a different approach more closely tailored to local circumstances may be needed. This second approach, known as specific adaptation, ensures varieties that are more closely suited to farmers' needs.

Farmers are not interested in varieties adapted to a wide range of environments, but want germplasm tailored to the specific set of constraints they experience. It is difficult to explain to them that a variety that performs well under their own farming conditions will not be made available through the formal system because it is insufficiently broadly adapted.

A further advantage of specific adaptation is that it favours increased diversity in farmers' fields, whereas broad adaptation may reduce the genetic base, creating a dangerous vulnerability to new pests and diseases or to extreme climatic fluctuations.

## Selection criteria

Despite improved diagnostic research and the increased client orientation of many breeding programmes in recent years, differences may still remain between farmers' and breeders' selection criteria. As explained above, these differences often arise at local level, due to the prevalence of local stresses or quality characteristics. Listed below are some of the criteria likely to cause problems in this respect:

- resistance to local strains of pests and diseases,
- response to local climatic and soil conditions,
- taste/palatability,
- suitability for local processing,
- ritual functions,
- storage characteristics,
- use of byproducts,
- labour requirements (gender-specific),
- place in cropping cycle.

## Uniformity and diversity

When modern plant breeding began in the last century, the importance of diversity in farmers' fields as a cushion against fluctuating environmental conditions was little understood. Scientists aimed to develop crop varieties suited to monoculture. This approach was still prevalent in the early 1950s and 60s, when formal research on food crops in developing countries began and the early Green Revolution varieties were developed. These varieties, mostly bred to deal with the relatively homogeneous set of environmental conditions obtaining in irrigated agriculture, are widely thought to have narrowed the genetic base of the world's important food staples.

As scientists studied the cropping systems used by resource-poor farmers, the value of diversity, both between and within species, began to be appreciated. Although further progress in this direction needs to be made, breeding efforts have now started to focus more on the needs of intercropping or rotational systems, moving away from the monocropping approaches of the past. While old attitudes prevail in some quarters, the aim of the younger generation of plant breeders, especially those in the second generation of international research centres set up during the 1970s and 80s, is to increase the choice of germplasm available to farmers, rather than to secure the adoption of a single variety over large areas.

Despite this change in thinking, the combined issues of genetic uniformity and the requirement for impact still tend to place the formal research system in a "no-win" situation. If an improved variety has demonstrated superiority and is widely adopted by farmers, the system stands accused of reducing diversity; if not, then it is seen as lacking in effectiveness. SADC/ICRISAT (1994) reports that the recent widespread adoption of new, short-duration sorghum in Zimbabwe once again raises "Green Revolution-type concerns" that diversity is being lost. Yet the breeding programme concerned belongs to the CGIAR, widely criticised even today for its apparent lack of impact in Africa.

The debate on genetic uniformity often fails to reflect the complexity of the issue. Diversity operates at different hierarchical levels of the agricultural system, from land use, through the mix of crop species, to the phenology of different varieties and, ultimately, to the genetic level. Plants that look alike can be quite different genetically, and the reverse may also be true. In *Phaseolus* bean, for example, characters such as seed colour and habit are controlled by very few genes, a tiny proportion of the total genome. This means that the apparent diversity seen in farmers' fields may be deceptive. Moreover, it is important to distinguish between the reduction of genetic diversity and its total loss. Reduction may occur locally, when a new variety gains acceptance and displaces landraces, but seen over a wider geographical area the net balance of diversity may in fact be positive, by virtue of the new genes introduced.

Using genetic markers, scientists are now able to penetrate beneath the crop's phenotype to explore diversity at the genotypic level. Studies of this kind demonstrate the potential of modern biotechnology to increase scientists' understanding and use of genetic diversity.

In short, the modern, global system of germplasm exchange and genetic enhancement can be used to increase diversity rather than to reduce it. In future, we may expect the formal system to attach increasing importance to diversity, counteracting the alleged tendency of the past 40 years to narrow the genetic base of major food staples.

## Advent of advanced research techniques

The idea, widely put about during the early 1980s, that modern biotechnology would revolutionise plant breeding, turned out to be exaggerated (Hardon, 1991). Even in the industrialised countries, the advanced research methods developed in recent years have, with a few notable exceptions, so far had little impact on crop improvement. However, the disappointments of the 1980s have led to a more sober assessment of prospects and a better understanding of those applications likely to prove useful in tackling specific problems.

The potential of modern biotechnology to address the crop improvement needs of LEIA systems in developing countries thus appears more limited than it did a decade ago. Nevertheless, the experiences gained by the international agricultural research centres and their partners have revealed several areas of interest, and a number of new biotechnology-based innovations are now nearing on-farm testing in resource-poor farmers' fields. These are outlined in Chapter 7.

## Varietal dissemination

The dissemination of improved varieties is often characterised by a vertically organised production and extension system, in which certified seed of officially approved varieties is released under conditions of strict quality control. "Leakages" in such systems are increasingly reported, as farmers eager to obtain new seeds and scientists eager to disseminate them conspire, often with the connivance of NGOs or even of sympathetic government officials, to circumvent official procedures. For instance, in India several rice varieties rejected in official testing procedures have now become widely available to farmers (Maurya et al., 1988). Although it safeguards seed health and ensures that new materials are suited to farmers' needs, the formal seed system is proving too slow to meet demand.

Even when new varieties are released according to official procedures, they may be quickly absorbed within the informal seed selection system. Examples are *Phaseolus* beans in Rwanda (Sperling et al., 1993) and hybrid maize varieties in Mexico (Louette, 1994) and Central America (Almekinders et al., 1994).

In the formal public system, poor links between technology generation and technology transfer, including the functions of seed multiplication and dissemination, are one of the reasons why the overall adoption rates for improved varieties remain low. It is estimated that the share of pure seed from the formal sector rarely exceeds 10% for most staple food crops. Commodities in which hybrid seeds are in strong demand, such as maize and sorghum, are the main exceptions (Almekinders et al., 1994). Private seed companies remain unwilling to market the seed of open-pollinated varieties because of the unpredictability of demand.

Increasingly, public-sector research institutes are turning to NGOs and farmers' organisations to ensure effective dissemination. Much of the collaboration at present is with the larger, international NGOs, and comes as part of relief efforts in response to emergencies. In Rwanda, eight international agricultural research centres worked with national research systems and NGOs in the Seeds of Hope project, designed to help agriculture recover and to prevent the loss of valuable genetic diversity in the wake of the civil war there (CIAT, 1995). In Mozambique, World Vision embarked on the large-scale dissemination of two improved sorghum varieties developed by the International Crops Research Institute for the Semi-Arid Tropics (ICRISAT). The seeds were multiplied and distributed to farmers settling down to cultivate after 16 years of civil war (ICRISAT, 1995).

Besides these high-profile responses to short-term humanitarian needs, there are signs of less conspicuous but perhaps even

Ethiopian farmer showing three selected sorghum heads. Farmers with better harvests are often particularly clever seed selectors. Community seed banks, proposed by the Relief Society of Tigray (REST), enable the benefits of traditional seed selection to be extended to the whole community.
(*Trygve Berg*).

more useful longer-term relationships being established. In Kenya, for example, local, national and international NGOs regularly organise visits by farmers to research stations and on-farm trials, and are playing a growing role in the dissemination of improved varieties of several crops, especially pigeonpea. Similar levels of activity are reported in several other African countries.

## Integrating the two systems

Various authors have argued that local crop conservation and development should be more closely integrated with formal development efforts (Altieri and Merrick, 1987; Brush, 1991; Hardon and de Boef, 1993). These efforts would seek to support and enhance, rather than replace, community management of plant genetic resources. They would also extend at national level successful models developed locally.

The *baito* system in Tigray, Ethiopia, provides an example of this process at work in the field of conservation (see Box 6.5). This and similar experiences elsewhere have given rise to the concept of community-based conservation as potentially the most cost-effective way of storing germplasm.

---

**Box 6.5**
**Seed banks in Tigray**

In the course of discussions following the famine of 1984-85, rural people in the Tigray Province of Ethiopia complained that drought victims had been exploited by seed lenders who had charged exorbitant interest rates to those forced to obtain seed on credit in order to start farming again. People had also noticed that some farmers had better harvests than others in spite of similar climatic conditions, soils and crop husbandry practices. They concluded that the differences could be explained by the quality of the seeds used: the farmers with the better harvests were known to be particularly clever seed selectors.

These problems and perceptions came to the attention of the Relief Society of Tigray (REST), which proposed the establishment of community seed banks as a solution. The banks would enable the benefits of traditional seed selection to be extended to the whole farming community. They would also provide credit for seeds on non-exploitative terms.

By 1991, seed banks were operational in 42 *woredas* (sub-districts), covering a major part of Tigray. The banks are owned and managed by the *baito*, a traditional elected body at the *woreda* level. The *baito* identifies local experts in seed selection and purchases good seeds on their advice. The seeds are kept by trusted female seed keepers. Loans are granted by the *baito* to farmers, who receive seeds at planting time for the price paid by the *baito*. The borrowers repay the loan in cash with 6-9% interest after harvest. In cases of crop failure the *baito* may accept a year's delay in repayment without charging additional interest. There is strong social pressure to repay promptly, and so far most of the seed banks have managed to maintain their initial capital fairly well. Seed selectors and keepers provide their services free of charge. Any administrative costs are shouldered by the *baito* (Berg, 1992).

The community seedbank system is now being introduced in other regions of Ethiopia by the Ethiopian Gene Bank, working in partnership with local NGOs (Mekbib et al., 1993).

Source: Berg (1992) and Mekbib et al., (1993).

The Community Biodiversity Development and Conservation Programme, funded by a consortium of Western donors, is the first attempt to explore this concept at international level (CLADES et al., 1994). The aim of the programme is to conserve not so much farmers' crops themselves but the process by which farmers generate crop genetic diversity. Farmers and their local communities are left entirely free to choose which crops or varieties they wish to conserve and develop. *In situ* conservation "without the fence" is one way of describing such efforts. Now active in 30 crops scattered over 17 developing countries, the programme is seen as a complement to *ex situ* conservation in genebanks, not a replacement for it. If *in situ* conservation is relaxed to the point at which farmers are free to lose old land races if they wish to do so, then the "freezing" ability of genebanks comes in useful as a fallback.

Provided the local community is accepted as the starting point, there are plenty of situations in which technical assistance could be used to improve the quality of the seeds selected by farmers, to enhance farmers' storage methods, or to strengthen local capacities in seed production and distribution (Rhoades and Booth, 1982; Graham and Hesse, 1991).

Enabling breeders to build on a long period of adaptation to local stresses, land races have already proved extremely useful in breeding programmes conducted for marginal areas (Ceccarelli, 1994). The challenge here is to link farmers' knowledge and descriptions of such material with scientific characterisation by the formal sector. As we have seen, few plant collection missions currently allow time for gathering any but the most rudimentary information about accessions. Participatory seed identification and description exercises should therefore precede characterisation by scientists.

*All discussions on Plant Breeders' Rights need to take into account the fact that farmers are breeders.* Vandana Shiva, at the Conference of Parties to the Convention on Biological Diversity, Jakarta, 1995.

New bio- and information technology provide further opportunities for improving the efficiency and accuracy of characterisation by the formal sector. Following characterisation, the resulting information, together with the duplicated genetic materials, should be returned to the local community. This seldom happens at present. In the spirit of the 1992 Convention on Biodiversity, access to the germplasm originating in their fields should be guaranteed to farmers.

Turning to crop development, the innovative capacity of farmers in this area is seldom harnessed by formal breeding programmes at present. Sometimes it is not even recognised. Box 6.6 provides an example, describing a farmer innovation directly at variance with the modern emphasis on "pure" seed.

Various ways of strengthening the interaction between researchers and farmers so as to build on the latters' capacity for research have been proposed. Biggs (1988) classified approaches to on-farm research as being either contractual, consultative, collaborative or collegiate (Table 6.1). The collaborative mode has the advantage of obtaining farmers' reactions to prototype

**Box 6.6**
**Introgression in farmers'**
**sorghum varieties in**
**Zimbabwe**

Small-scale farmers in Zimbabwe are keen to exchange and experiment with new seed material. On-farm selection practices are innovative rather than conservative.

In one area, weedy sorghum is allowed to mature in cultivated fields while wild sorghum grows freely along field edges and on fallow land. Hybrids between cultivated and weedy and/or wild sorghum are often found in farmers' fields. From a conventional agronomic point of view, this is undesirable, since the cultivated gene pool becomes "contaminated". Yet this is potentially an important way of acquiring new genes. Traits such as disease resistance and drought tolerance may be slowly incorporated into the cultivated crop in this manner. Farmers control the introgression process by roguing off-types.

The biological potential of introgression in sorghum remains to be recognised by many sorghum plant breeders.

Source: Van Oosterhout (1993).

technologies early in the technology development process, while in the collegiate mode, farmers take the lead in technology design and development. Following this classification, Amanor et al. (1993) have suggested that more plant breeding research both could and should be conducted in a collegiate mode. Too many breeding programmes still follow a contractual mode, in which farmers are considered to be passive "technicians" whose role is merely to supply land and cheap labour for the "scientist managers" (Maurya et al., 1988; Chambers, 1993).

Several attempts have been made in recent years to increase the influence of farmers on the design of new technology. For instance, the Sorghum and Millet Improvement Program (SMIP) of the International Crops Research Institute for the Semi-Arid Tropics (ICRISAT), working with its national partners in Zimbabwe, has built nurseries representing the morphological diversity of sorghum genotypes found in the country. Groups of farmers are regularly invited to the research station and asked to identify their preferred genotypes, giving their reasons. The results are subjected to detailed analysis, after which they are fed back into the international and national breeding programmes as a corrective to scientists' breeding objectives (ICRISAT, 1995).

Though praiseworthy, these changes fall short of the wholesale reform of the formal system that some observers believe to be necessary if integration with the informal system is to succeed. In a model that relied more on farmers' research capacities, plant breeders could apply the techniques of the formal system to conduct strategic research on key problems identified by farmers, who would then take over the rest of the research process. Plant breeders might disseminate segregating populations, leaving part of the selection process to farmers. Farmers could be taught how to recombine genotypes (varietal crossing) and how to apply selection methods other than mass selection. Table 6.2 illustrates the range of options available in farmer, institutional and shared control of the various steps in plant breeding.

TABLE 6.1    On-farm research modes

| Mode | Objective |
| --- | --- |
| Contractual | Scientists contract with farmers to provide land or services |
| Consultative | Scientists consult farmers about their problems and then develop solutions |
| Collaborative | Scientists and farmers collaborate as partners in the research process |
| Collegiate | Scientists work to strengthen farmers' informal research and development systems in rural areas. |

Source: Biggs (1988).

TABLE 6.2    The range of options in control of different steps of the plant breeding process

| Breeding cycle | Farmers | Breeders |
| --- | --- | --- |
| *Setting breeding objectives* | ++ | ++ |
| *Germplasm enhancement* | +++ | +++ |
| *Identify variation* | | |
| Identify sources of variation (evaluation): | | |
|     specific adaptation | ++ | + |
|     resistance | + | ++ |
| Recombination (crossing) | | |
|     self-pollinating crops | – | ++ |
|     cross-pollinating crops | + | +++ |
|     vegetatively propagated crops | – | +++ |
| *Selection* | | |
|     segregating populations: | | |
|         early selection (F2/F3) | + | +++ |
|         mid selection (F4/F5) | ++ | +++ |
|     populations of lines | +++ | ++ |
|     finished lines/varieties | +++ | + |
| *Release and distribution* | +++ | ++ |

Adapted from Hardon (1995).
– till +++ are indicators for an increased importance of the control by the farmer or professional breeder in the breeding process.

Working with varying degrees of institutional support, several researchers in the formal system are currently trying out a range of approaches to participatory plant breeding. Table 6.3 summarises these approaches and their products.

TABLE 6.3 Approaches in participatory plant breeding

| Institute | Country | Crop | Reproduction system | Sources of variation | Selection | | | Output |
|---|---|---|---|---|---|---|---|---|
| | | | | | What is selected? | Where selection takes place? | Who controls/performs selection? | |
| ICARDA | Syria | barley | self-pollinator | local landraces | lines | on-farm | researcher-managed | pure lines of landraces |
| | Syria | barley | self-pollinator | local landraces other landraces | segregating populations | on-station | researcher-managed | advanced populations for further breeding |
| CIAT | Rwanda | common beans | self-pollinator | local landraces modern varieties | lines | on-station | farmer-evaluated | many lines in mixtures; farmers' selection criteria |
| | Colombia | common beans | self-pollinator | landraces modern varieties | segregating populations | on-station on-farm | farmer-evaluated farmer-evaluated | advanced lines for further use in breeding process |
| EMBRAPA | Brazil | common beans | self-pollinator | local landraces other landraces modern varieties | segregating populations | on-station & on-farm | farmer-evaluated | farmers' selection criteria; advanced population; informal distribution |
| CIP | Peru | potato & other tubers | vegetatively propagated | local landraces & modern varieties | clones | on-farm | farmer-managed | identification of clones in mixtures |
| ICRISAT | India | pearl millet | cross-pollinator | local landraces other landraces modern varieties | populations | on-farm & on-station | farmer evaluated | improved populations for further breeding |
| LARC | Nepal | rice | self-pollinator | landraces modern varieties | segregating populations | on-farm | researcher-managed | farmers' selection criteria; informal distribution |
| KRIBP | India | maize | cross-pollinated | landraces modern varieties | composites | on-farm | farmer-evaluated | advanced populations |
| | India | various | – | modern varieties (released) | varieties | on-farm | farmer-evaluated | identify appropriate modern varieties |
| IRRI | India | rice | self-pollinated | local landraces other landraces modern varieties | lines | on-farm | farmer/researcher-managed | advanced lines for further use in the breeding process |

Adapted from Hardon (1995).

Low-cost tissue culture, used to generate virus-free materials, is one modern biotechnology that can be used to support farmers' seed multiplication efforts. Such technology is particularly applicable to vegetatively propagated crops such as potato, cassava and yam (see Chapter 7).

# Conclusion

The challenge facing farmers' crop conservation and development is to find ways of enhancing local knowledge and linking it to world science, while at the same time ensuring that control of that knowledge remains within the community. This challenge requires researchers to address the policy dimensions of their work, so as to promote reform in such areas as intellectual property rights. At the same time, local organisations need to be strengthened so as to increase the capacity of farmers to manage local conservation and development, analyse the wider world in which their efforts unfold, and exercise effective pressure on institutions and policy makers in the formal sector.

## References

Allard, R.W. 1990, 'The genetics of host-pathogen co-evolution: Implications for genetic resource conservation,' *Journal of Heredity* 81 (1): 1–6

Almekinders, C.J.M., Louwaars, N.P. and de Bruijn, G.H. 1994, 'Local seed systems and their importance for an improved seed supply in developing countries', *Euphytica* 78: 207–216

Altieri, M.A. and Merrick, L.C. 1987, 'In-situ conservation of crop genetic resources through maintenance of traditional farming systems,' *Economic Botany* 41 (1): 86–96

Amanor, K., Wellard, K., de Boef, W. and Bebbington, A. 1993, 'Cultivating knowledge: Genetic diversity, farmer experimentation and crop research,' in de Boef, W., Amanor, K., Wellard, K., with Bebbington, A. (eds), *Cultivating Knowledge: Genetic Diversity, Farmer Experimentation and Crop Research*, Intermediate Technology Publications, London: 1–13

Benzing, A., 1989, 'Andean potato peasants are seed bankers,' *ILEIA Newsletter* 5 (4) : 12–14

Berg, T., 1992, 'Indigenous knowledge and plant breeding in Tigray, Ethiopia,' *Forum for Development Studies* 1: 13–22

Berg, T., 1995, *Dynamic Management of Plant Genetic Resources: Potentials of Emerging Grass Roots Movements*, Food and Agriculture Organisation, Rome

Berg, T. and Alcid, M.L. 1994, Review of Community-based Native Seed Research Center (CONSERVE), Mindanao, Philippines, unpublished report to the Development Fund, Oslo, Norway

Berg, T., Bjoernstad, A., Fowler, C. and Kroeppa, T. 1991, 'Technology Options and the Gene Struggle,' Occasional Papers Series C, Development and Environment No. 8, Agriculture University of Norway, Norwegian Centre for International Agricultural Development (NORAGRIC), Ås, Norway

Biggs, S.D., 1988, 'Resource-poor farmer participation in research: A synthesis of experiences in nine national research systems,' On-Farm Client-

Oriented Research Comparative Study Paper No. 3, International Service for National Agricultural Research, The Hague, Netherlands

Brown, A.H.D. 1978, 'Isozymes, plant population genetic structure, and genetic conservation,' *Theoretical and Applied Genetics* 52: 145–157

Brown, A.H.D. 1979, 'Enzyme polymorphism in plant populations,' *Population Biology* 15: 1–42

Brush, S.B. 1991, 'A farmer-based approach to conserving crop germplasm,' *Economic Botany* 45 (2): 153–165

Ceccarelli, S. 1989, 'Wide adaptation: How wide?' *Euphytica* 40: 197–205

Ceccarelli, S. 1994, 'Specific adaptation and breeding for marginal conditions,' *Euphytica* 77 (3): 205–219

Ceccarelli, S., Valkoun, J., Erskine, W., Weigand, S., Miller, R. and van Leur, J.A.G. 1992, 'Plant genetic resources and plant improvement as tools to develop sustainable agriculture,' *Experimental Agriculture* 28: 89–98

Chambers, R. 1993, *Challenging the Professions: Frontiers for Rural Development,* Intermediate Technology Publications, London

CIAT. 1995, 'CIAT in perspective, Annual Report 1994, Centro Internacional de Agricultura Tropical, Cali, Colombia

CLADES, COMMUTECH, CPRO-DLO, GRAIN, NORAGRIC, PGRC/E, RAFI and SEARICE. 1994 Community Biodiversity Development and Conservation Programme: Proposal to DGIS, IDRC and SIDA for Implementation, Phase I: 1994–1997, CPRO-DLO Centre for Genetic Resources (CGN) and Centro de Educación y Tecnología (CET), Wageningen, Netherlands and Santiago, Chile

Coffman, W.R. and Smith, M.E. 1991, 'Roles of public, industry, and international research centre programmes in developing germplasm for sustainable agriculture,' in Sleper, D.A., Barker, T.C., Bramel-Cox, P.J. and Francis, C.A. (eds), *Plant Breeding and Sustainable Agriculture: Considerations for Objectives and Methods*, Proceedings of a Symposium in Las Vegas, CSSA Publication No. 18: 1–9, Crop Science Society of America, Madison

Crucible Group, 1994, *People, Plants and Patents: The Impact of Intellectual Property on Biodiversity, Conservation, Trade and Rural Society,* International Development Research Centre, Ottawa, USA

De Boef. W., Amanor, K., Wellard, K., with Bebbington, A. 1993, *Cultivating Knowledge: Genetic Diversity, Farmer Experimentation and Crop Research,* Intermediate Technology Publications, London

Fowler, C. 1994 *Unnatural Selection: Technology, Politics and Plant Evolution.* International Studies in Global Change, No. 6. Gordon and Breach, Yverdon, Switzerland

Graham, O. and Hesse, C. 1991, *Cereal Banks: At your Service? The Story of Toundeu-Patar, A Village Somewhere in the Sahel,* OXFAM publication on behalf of the Arid Lands Information Network, Oxford, UK

Hardon, J.J. 1991, 'Biotechnology, plant breeding and resource-poor farmers in the Third World,' paper presented at the Public Debate on Biotechnology and Farmers' Rights: Opportunities and Threats for Small-scale Farmers in Developing Countries, 8–9 April, 1991, Amsterdam

Hardon, J.J. 1995, 'Participatory plant breeding: Outcome of a workshop sponsored by IDRC, IPGRI, FAO and CGN at Wageningen, Netherlands, 26–29 July 1995,' Issues in Genetic Resources No. 3. International Plant Genetic Resources Institute, Rome

Hardon, J.J. and de Boef, W.S. 1993, 'Linking farmers and breeders in local crop development,' in de Boef, W., Amanor, K., Wellard, K., with Bebbington, A. (eds), *Cultivating Knowledge: Genetic Diversity, Farmer*

*Experimentation and Crop Research,* Intermediate Technology Publications, London

Hodgkin, T., Ramanatha Rao, V. and Riley, K. 1993 'Current issues in conserving crop landraces *in situ,'* paper presented at the On-Farm Conservation Workshop, 6–8 December 1993, Bogor, Indonesia

ICRISAT. 1995, Southern and Eastern Africa Regional Programme Annual Report 1994, ICRISAT, Bulawayo, Zimbabwe

Janssen, W., Luna, C.A. and Duque, M.C. 1992, 'Small farmer behaviour towards bean seed: Evidence from Colombia,' *Journal of Applied Seed Production* 10: 43–51

Keystone Center. 1991. Global Initiative for the Security and Sustainable Use of Plant Genetic Resources: Final Consensus Report of the Keystone International Dialogue Series on Plant Genetic Resources, Third Plenary Session, 31 May–4 June 1991, Oslo, Norway. Keystone Center, Keystone, Colorado, USA

Linnemann, A.R. and Siemonsma, J.S. 1989, 'Variety choice and seed supply by smallholders,' *ILEIA Newsletter* 89 (4): 22–23

Longley, C. and Richards, P. 1993, 'Selection strategies of rice farmers in Sierra Leone,' in de Boef, W., Amanor, K., Wellard, K., with Bebbington, A. (eds), *Cultivating Knowledge: Genetic Diversity, Farmer Experimentation and Crop Research,* Intermediate Technology Publications, London

Louette, D. 1994, Gestion Traditionnelle de Variétés de Maïs dans la Réserve de la Biosphère Sierra de Manantlàn (RBSM, Etats de Jalisco et Colima, Mexique) et Conservation *in situ* de Ressources Génétiques de Plantes Cultivées, Thèse de doctorat, ENSAM, Montpellier, France

Maurya, D.H., Bottral, A. and Farrington, J. 1988, 'Improved livelihoods, genetic diversity and farmer participation: A strategy for rice breeding in rainfed areas in India,' *Experimental Agriculture* 24 (3): 311–320

McNeely. J. 1989, "Conserving genetic resources at farm level," *ILEIA Newsletter* 4/89

Mekbib, H., Mariam, B.G. and Eyasu, H.S. 1993, Technical Consultancy Service on Relief Society of Tigray (REST) Seed Bank Programme in Tigray, PGRC/E, Addis Ababa

National Research Council. 1989, *Lost Crops of the Incas: Little-known Plants of the Andes with Promise for Worldwide Cultivation,* National Academy Press, Washington DC, USA

Osman, A.M. 1990, 'Changing Ifugao agriculture: Roles of cultivators and agro-technologists in developing terrace rice cultivation in the Philippines,' M.Sc. thesis, Department of Rural Sociology of Developing Countries, Wageningen Agricultural University, Netherlands

Pereira, W. 1990, 'The sustainable lifestyle of the Warli,' in Pereira, W. and Seabrook, J. (eds), *Asking the Earth,* Earthscan, London

Pingali, P.L. 1993, 'Opportunities for the diversification of Asian rice farming systems: A deterministic paradigm,' in Miranda, S.M. and Maglinao, A.R. (eds), *Irrigation Management for Rice-based Farming Systems in Bangladesh, Indonesia and the Philippines: Proceedings of a Workshop held in Colombo, 1990,* International Irrigation Management Institute

Prain, G.D. and Scheidegger, U. 1988, 'User-friendly Seed Programmes,' report of the Third Social Science Planning Conference held at CIP, Lima, Peru, 7–10 September 1987

Prescott-Allen, R. and Prescott-Allen, C. 1990, 'How many plants feed the world?' *Conservation Biology* 4 (4): 365–374

Rhoades, R.E. 1985, 'Traditional Potato Production and Farmers' Selection of Varieties in Eastern Nepal,' Potatoes in Food Systems Research Series No. 2, International Potato Centre (CIP), Lima

Rhoades, R.E. and Booth, R.H. 1982, 'Farmer-back-to-farmer: A model for generating acceptable agricultural technology,' *Agricultural Administration* 11: 127–137

SADC/ICRISAT, 1994, *Annual Progress Report of Work from 1 January to 31 December 1994*, Report submitted to the Bundesministerium für Wirtschaftliche Zusammenarbeit und Entwicklung and the Deutsche Gesellschaft für Technische Zusammenarbeit, International Crops Research Institute for the Semi-Arid Tropics, Hyderabad, India

Seetharam, A., Riley, K.W. and Harinarayana, G. 1990, 'Small Millets in global Agriculture,' proceedings of a workshop held in Bangalore, India, 1986, Aspect Publishing, London

Sperling, L., Loevinsohn, M.E. and Ntambovura, B. 1993, 'Rethinking the farmers' role in plant breeding: Local bean experts and on-station selection in Rwanda,' *Experimental Agriculture* 29: 509–519

Tapia, M.E. and Rosas, A. 1993, 'Seed fairs in the Andes: A strategy for local conservation of plant genetic resources,' in de Boef, W., Amanor, K., Wellard K., with Bebbington, A. (eds), *Cultivating Knowledge: Genetic Diversity, Farmer Experimentation and Crop Research,* Intermediate Technology Publications, London

Van der Maesen, L.G.J., Somaatmadja, S., Wulijarni-Soetjipto, N. and Siemonsma, J.S. 1989, *Pulses, Plant Resources of South-East Asia No. 1.* Pudoc, Wageningen, Netherlands

Van Oosterhout, S. 1993, 'Developing an *in situ* conservation programme for indigenous and traditional crops in Zimbabwe,' paper presented at the Regional Conference for the Conservation and Utilisation of Botanical Diversity in Southern Africa, Cape Town, South Africa

Weltzien, E. and Fischbeck, G. 1990, 'Performance and variability of local barley landraces in Near-Eastern environments,' *Plant Breeding* 104: 58–67

Wolf, E.C. 1986, *Beyond the Green Revolution: New Approaches for Third World Agriculture,* World Resources Institute, Washington DC, USA

Worede, M. and Mekbib, H. 1993, 'Linking genetic resource conservation to farmers in Ethiopia,' in de Boef, W., Amanor, K., Wellard, K., with Bebbington, A. (eds), *Cultivating Knowledge: Genetic Diversity, Farmer Experimentation and Crop Research,* Intermediate Technology Publications, London

# Part 2

---

# Science-based Biotechnology

# 7  Assessing the Potential

*Jacqueline E.W. Broerse and Bert Visser*

## Introduction

Can science-based biotechnology research benefit resource-poor farmers in developing countries? This question is especially important given the relatively high investment costs and sophisticated nature of such research, the capacity for which is generally much lower in the developing countries than in the developed world. This chapter will review the major biotechnologies so far developed through formal research, describing their current applications and the prerequisites for their successful introduction.

## Science-based biotechnology

We will restrict ourselves to a discussion of the following fields and applications:

- food processing: fermentation, enzyme technology and monoclonal antibody technology,
- plant nutrition and health: biofertilisers, biocontrol agents and diagnostics (immuno-assays and DNA probes),
- animal nutrition and health: feed (enzyme technology), digestion, metabolism (genetic modification); drug and vaccine development (fermentation technology, genetic modification) and diagnostics (immuno-assays and DNA probes),
- germplasm improvement and conservation: selection and breeding (tissue culture, genetic markers, genetic modification, mutation induction, somatic hybridisation and cybridisation); reproduction (mass propagation, detection of hormone levels, embryo technology); germplasm conservation (tissue culture, isozyme analysis).

As we have already seen, some of these technologies, notably fermentation and to a certain extent biofertilisers and biocontrol agents, were originally developed centuries ago, independently of the modern laboratory, and have been used and adapted by millions of domestic users down the generations. Most, however, involve molecular, cellular or tissue-level intervention and stem from laboratory research. Recent developments concern molecular intervention, notably the isolation, characterisation,

*The world of our making has become so complicated that we have to turn to the world of nature to understand it.* Kevin Kelly, Out of Control: The Rise of Neo-Biological Civilisation, 1995.

modification and transfer of genes, and the isolation and characterisation of DNA probes for use as genetic markers in plant and animal breeding. Applications of these technologies are only just beginning.

# Food processing

Science-based biotechnologies for food processing include fermentation technology, enzyme technology and monoclonal antibody technology.

## Fermentation technology

For thousands of years mankind has used micro-organisms to produce foods and drinks without understanding the often complex microbial processes underlying their production. It was not until the latter half of the 19th century that attempts were made to analyse and standardise these processes so as to improve product quality and uniformity.

The fermentation process can be divided into three phases (Greenshields and Rothman, 1986):

- Preparation. An organism with appropriate characteristics is selected. Subsequently, it is grown to provide an inoculum.
- Fermentation proper. This takes place in a fermentor or bioreactor, which may vary from a simple hole in the ground to a highly sophisticated, large-scale, computerised production unit. Two types of fermentation can be distinguished; (i) non-aseptic systems, as used in traditional processes, which do not need strict monocultures; and (ii) aseptic systems, of the kind applied to produce antibiotics, amino acids and single-cell proteins, from which all micro-organisms but the one of interest must be excluded. These systems are commonly used in modern fermentation processes.
- Harvesting and product recovery (downstream processing). Following the growth and production phases, the micro-organisms may have to be separated from the fermentation medium and product. This can be achieved by precipitation, filtration or centrifugation.

Although all fermentations require water, it is convenient to differentiate between fermentations in liquid medium and solid-state fermentations, where much less water is present.

### Liquid-medium fermentation

Liquid-medium fermentation encompasses alcoholic fermentation, lactic acid fermentation, industrial production of other microbial metabolites, and waste water treatment. Alcoholic and lactic acid fermentation are the main forms used in food fermentation.

Alcoholic fermentation usually takes the form of brewing, in which yeasts convert sugar to ethanol (alcohol) in the absence of oxygen. Ethanol is produced in many small-scale settings, but is also commercially important as the basis of an enormously varied beer industry and as a liquid fuel. Acetic acid bacteria also play an important role in alcoholic fermentations, converting alcohol to acetic acid. This process is used in the production of vinegar. Major problems in alcoholic fermentation are contamination of the culture with undesired bacteria and the over-production of acids.

A wide variety of raw materials, including cereals, milk, vegetables and fish, are subjected to the activity of lactic acid bacteria. Lactic acid fermentations are generally inexpensive and often require little or no heat. Their major problem is slow acid formation, which can result in off-flavours, extended process times or complete loss of the product.

## Solid-state fermentation

In solid-state fermentation (SSF) processes, microbial growth and product formation occur on the surfaces of solid substrates. Because of the heterogeneity of these substrates, SSF is not as well characterised and cannot be as easily controlled as liquid-medium fermentation. It is, however, widely used in developing countries, especially in South-East Asia. SSF includes a number of well-known microbial processes such as mushroom cultivation and the production of mould-ripened cheese, among many other familiar household products (see Chapter 5). Substrates traditionally fermented in solid state include rice, wheat, millet, barley, maize, soybean and cassava.

SSF is an unsophisticated technology; solid substrates require only the addition of a little water, and low moisture reduces the risk of contamination. Agricultural substrates may, however, require some kind of pre-treatment such as cracking or surface abrasion (Mudgett, 1986). The major technical problem with SSF is the difficulty of scaling up the process while avoiding heat build-up (Mudgett, 1986; Rai et al., 1988).

## Improvement of food fermentation

The efficiency and yield of food fermentation processes can be increased at all three stages (preparation, fermentation proper and harvesting) through the selection or development of more productive microbial strains, the control of culture conditions, or the improvement of product purification and concentration. Science-based biotechnological research focuses mainly on improving the metabolic properties of the micro-organisms used as a starter culture, using the techniques of selection, mutation and genetic modification. Several properties are the subject of genetic modification, among them the production of bacteriocins and acids. Bacteriocins, which inhibit the growth of competitors, are natural products of certain bacteria, including some strains which occur in food fermentations. The transfer of genes encoding bacteriocin production to microbial starter cultures can improve the safety of fermented foods by inhibiting the growth of pathogenic bacterial strains. Acid production by lactic acid bacteria can be enhanced or accelerated by increasing the number of genes encoding the enzymes responsible.

A prerequisite for this type of research is a full understanding of the fermentation process. The properties of the starter culture must be known and the culture conditions must be fully controllable. Consequently, science-based biotechnological research currently has little to offer indigenous food processing, in which fermentation is usually an uncontrolled process using a non-sterile substrate and a mixed starter culture of an unknown combination of micro-organisms.

Nonetheless, the need to improve traditional fermentation processes is generally acknowledged (see Chapter 6). The disadvantages of these processes are that the yield is often low and the product may not always be safe and is frequently of variable quality. To overcome these problems, a better understanding of the underlying microbial processes is needed. This would include: isolation and characterisation of the essential micro-organisms; determination of the way in which the physical and chemical environment affects the metabolism of these micro-organisms; investigation of the effects of pretreatment of raw material on the fermentation process; identification of the options for down-stream processing and how these affect the taste and texture of the product; in short, an improved understanding of the microbiology and biochemistry of the entire fermentation process. This type of research will be costly and time-consuming, since most of the underlying microbial processes in traditional food processing are not only unknown but also highly complex. Worldwide, few research groups conduct research on traditional food fermentation at present.

In most developing countries, research institutes and university laboratories have facilities for conducting research on fermentation, particularly for the identification and characterisation of micro-organisms. Both in developed and developing countries, genetic modification still plays a marginal role in food fermentations. However, there are many examples of domestic industries applying other fermentation techniques. Many developing countries have an extensive beer and wine industry, with modern large-scale facilities operating alongside the traditional sector.

Besides alcoholic fermentations, liquid lactic acid and solid-state fermentation processes are widely practised in developing countries. With some exceptions, traditional food fermentation processes operate at the domestic/village level rather than at the industrial level. There have been few successful attempts to scale up the production of indigenous fermented foods. Successful examples are the production of fermented ground cereal, cassava, palm wine, sorghum beer, carob and soy sauce on an (semi-) industrial scale (Okafor, 1992). There are several technical difficulties. It is almost impossible to imitate the rich and full flavours of the traditional product when using the pure starter cultures and aseptic conditions required for large-scale processing (Nout, 1992). Better control of fermentations operating under non-sterile conditions would be a more appropriate approach, but this is complicated and might not lead to satisfactory results. An example of these difficulties is given in Box 7.1.

An alternative to scaling up traditional fermentation processes is to improve them at the domestic and village levels. Despite improvements achieved through formal research, virtually no science-based biotechnologies relevant to food processing have reached production at these levels, usually because they involve sophisticated equipment with high investment and operating costs and would fail to translate into higher profits. So far, most innovations at village level are the result of the producers' own experimentation. One exception is cassava, a crop for which a range of new products and processes has been jointly developed by farmers and scientists.

## Enzyme technology

An enzyme is a protein that catalyses a specific chemical reaction. Unlike the agents responsible for most other chemical processes, enzymes normally operate in conditions of mild pH (four to eight), low temperature (10-80°C) and normal atmospheric pressure. Purified enzymes can be used to improve fermentation and other types of food processing. Enzyme technology can be divided into four phases (Towalski and Rothman, 1986):

- Extraction: Extracellular enzymes (secreted by the cells into the environment) can be processed directly from the culture medium. For intracellular enzymes (which catalyse reactions within the cell itself), cells need to be disrupted to free the enzymes. Surfactants are used to dislodge the enzymes from cell frag-

---

**Box 7.1**
**Fermented milk in Zimbabwe**

In Zimbabwe, the rural population ferments its milk traditionally by allowing it to stand at ambient temperature in an earthenware pot loosely covered by a plate. Fermentation is caused by micro-organisms present in the milk, pot and surrounding air, and takes one to two days. The fermented milk has a shelf life of three days.

Although an industrially produced fermented milk product, Lacto, is available, most people prefer the taste of the traditional product. Analysis of both products by researchers at the University of Zimbabwe showed that the traditional one contained a much wider variety of micro-organisms and more vitamins and flavour compounds. The researchers concluded that the traditional process was extremely complex and that it would be very difficult and time-consuming to prepare an appropriate starter culture that would result in a manufactured product that tasted as good as the traditional one.

Source: Feresu (1992).

ments. The extracts are usually very dilute and may contain many impurities.

- Purification and concentration: The extract is filtrated or centrifuged, and enzymes are precipitated or are selectively bound to specific carriers and later eluted.
- Stabilisation: The stability of enzymes is improved by mixing them with other proteins, salts, starch hydrolysates or sugar alcohols.
- Preparation: Enzymes can be prepared either; (i) in liquid form, as crude extracts or as highly purified and stabilised preparations; or (ii) in solid form as pure crystals, freeze-dried (semi-pure), or immobilised in or on a wide range of carriers.

The use of purified enzymes is over 50 years old. Although several thousands of enzymes have been characterised, only about 350 are used, with a dozen or so accounting for the bulk of those commercially available. All enzymes are obtained from living organisms, usually microbes but sometimes animal organs or plant tissues. The utilisation of enzymes has been improved considerably since the advent of immobilisation techniques, by which enzymes can be attached to solid surfaces rather than having to be maintained in solution. This makes them re-usable and opens up the way to continuous processing. Today, enzymes have five distinct areas of application: in cosmetics, therapy, the food and feed industry, for diagnostic purposes, and in research.

One of the most important recent applications is the production of foodstuffs from non-traditional raw materials. An example is the development of a sweetener named high fructose corn syrup (HFCS), also called isoglucose. This innovation did little to benefit developing countries (Box 7.2). Another recent application is the use of enzymes in animal feed, for example the enzyme phytase (see Box 7.3, p. 150).

---

**Box 7.2**
**Enzyme technology and sweeteners**

In 1965, Japanese biotechnologists identified a bacterium which produces high yields of the enzyme glucose isomerase. The isolation of this enzyme opened the way to the commercial production of isoglucose from starch-rich crops such as maize, wheat and potato.

Following the introduction of isoglucose in soft drinks, sugar consumption in the USA fell from 10.8 million tonnes in 1979 to 8.6 million tonnes in 1984. In 1984, the Coca Cola Company and Pepsico disclosed their intention to substitute all sugar by isoglucose in soft drinks within two years. The market price of raw sugar collapsed from around US$ 0.62 in 1980 to only US$ 0.12 in 1985. Neither company offered any compensation or subsidy to their traditional suppliers that would have enabled them to convert from the production of sugar cane to that of other crops.

Farmers in developing countries cannot easily cope with such rapid changes in market demand. The impact of this biotechnological innovation was worst in those countries whose economy depended greatly on the export of sugar cane, such as the Philippines and the islands of the Carribean.

Sources: Hobbelink (1991), Ruivenkamp (1989).

Enzymes may be produced at relatively low cost in virtually unlimited quantities. Their application allows large savings in the fixed capital and operating costs of certain industries, since enzyme-catalysed processes operate under milder conditions than their chemical counterparts. The main disadvantage of enzymes is their instability; they are susceptible to denaturation and inhibition when their physical environment alters even only slightly, and their action is easily blocked by chemicals (Antebi and Fishlock, 1985; Towalski and Rothman, 1986). To counter this, genetic modification is being used to increase the stability and activity of enzymes. Another complication with some enzymes is that they require co-factors; thermostable, non-protein molecules that cooperate with the enzyme proper to facilitate the catalytic reaction. Co-factors are usually costly to produce.

The production, purification and immobilisation of enzymes is complicated. Most developing countries cannot yet produce high-quality enzymes, but a few technologically more advanced ones, including Thailand, India, Brazil, Cuba and Mexico, are active in this area. The utilisation of enzymes in food processing is, in contrast, easy and widespread. For the economies of developing countries, enzyme technology presents both opportunities (import substitution through the increased use of local raw materials) and risks (loss of export markets, see for example Box 7.2). The use of enzymes in food processing is likely to continue to apply far more to large-scale industrial food processing than to small-scale operations.

## Monoclonal antibody technology

In the monoclonal antibody technology, an antibody-producing cell (1) is fused with a cancer cell. The antibody-producing cell brings the ability to produce a particular type of antibody, while the cancer cell endows the hybrid with immortality in culture. The resultant hybridoma cell, and all cells derived from it by cell division, will produce large amounts of a single antibody, a monoclonal antibody (MAb) (Baker, 1989; Stribley, 1989). Hybridoma cells can be grown in large fermentors or immobilised on solid surfaces. The antibodies are secreted into the growth medium, which allows easy harvesting. The first report of hybridoma production was in 1975. MAb production is now widely used commercially.

Monoclonal antibodies are mainly used for diagnostic purposes, to identify and quantify pathogenic organisms (viruses, bacteria and parasites), tumours, hormones, metabolites, drugs and toxins; if they are present in plants, (animal) body fluids, tissues or foodstuffs. For antigen-antibody binding to be detected and quantified, the antibodies must be labelled with a marker molecule, usually an enzyme which catalyses a specific reaction resulting in an easily identifiable product (enzyme-linked immuno-sorbent assay, or ELISA). The advantage of MAbs in diagnostic immuno-assays is their relative speed, accuracy and specificity. As standard reagents, MAbs are undoubtedly superior to polyclonal antibodies, being readily available for an indefinite period at standard quality (Persley, 1990; Primrose, 1987; UNDP, 1989). MAbs can also be used in therapeutic treatment, and in protein purification.

Diagnostic tests are increasingly used in the quality control of foodstuffs. For example, Skerrit et al. (1993) indicate that diagnostic tests have been developed to predict dough properties in wheat, for measuring the gluten contents of cereal foods, for monitoring additives and fungal contaminants in brewing, and for contaminant detection in grains and cereal foods. Many of these diagnostic tests are ELISAs based on monoclonal antibodies.

Advances in cell fusion and genetic modification have led to the development of a second generation of monoclonal antibodies (Williams, 1988):

- Bi-specific monoclonal antibodies, in which the combining sites react with two distinct antigenic determinants and which can thus bind two distinct antigens.
- Chimaeric antibodies, in which the antigen-binding site (variable region) of the antibody, responsible for recognition, is from a rodent source but is genetically combined with a human constant region, thus combining the flexibility of selection in rodents with the retention of the human character of the MAb.
- Antibodies with novel effector functions: genes for proteins other than antibodies (e.g. protein toxins) can be spliced to genes for antibody fragments and expressed as a conjugate molecule.
- Single-chain antibodies, which are produced by genetically linking heavy and light chain variable region genes through a DNA sequence encoding a synthetic peptide linker. This allows more efficient production of these antibodies in bacteria and certain viruses.

These novel types of antibody have so far been intended mainly for use in human health care, to increase specificity and reduce anti-antibody responses (allergic reactions).

Although the technology is well developed, relatively robust and simple, few laboratories in developing countries can yet develop and/or produce MAbs. The feasibility of the technology, combined with its obvious relevance for developing countries, not only in food processing but also in plant and animal health care, mark it out as a candidate for support.

The range of MAb diagnostic tests available is likely to grow rapidly in the future, offering a wide variety of accurate, robust and easy-to-use kits applicable to many practical problems. For the time being, developing countries must purchase most of these kits abroad, but in the longer term they could develop them nationally. The facilities and investment costs needed for large-scale production are quite extensive, but production could start on a modest scale with simple equipment, using cheap labour.

Investment costs, at around US$ 50 000–100 000 for a functioning small-scale laboratory, are modest.

MAb development and production is, however, more labour- and cost-intensive than the conventional production of polyclonal antisera. It may take from a few months to a few years to obtain a desired MAb, depending on the expertise of the scientist, the quality of the facilities and the characteristics of the antigen to be isolated. For this reason, laboratories in developed nations commonly exchange hybridoma cells. Although laboratories in developing countries could, in addition to organising their own exchanges, purchase hybridoma cells, in practice this rarely occurs due to communication, logistic and financial problems.

The relatively high cost of MAb diagnostic tests mitigates against their application to food production or processing problems at domestic or village level. Their use will probably be limited to governmental food quality control laboratories and to some local large-scale food processing industries for the foreseeable future.

## Plant nutrition and health

Crop yields in developing countries are reduced by a wide range of biotic and abiotic stresses. Micro-organisms and plants are closely interdependent, often entering into a mutually beneficial relationship that, among other things, enables specific stresses to be controlled or attenuated. Human beings have endeavoured to enhance these effects using microbial inoculants. These usually consist of bacteria suspended in some sterile carrier. The two main uses of microbial inoculants are as biofertilisers and as biocontrol agents to combat insect pests, weeds or diseases. Science-based biotechnology plays a role in the development of both these technologies.

Another important area in which science-based biotechnology has already proved useful is the diagnosis of crop diseases. Simple and rapid diagnosis may assist in bringing about faster and more effective control, especially of viral diseases.

### Biofertilisers

Many micro-organisms influence nutrient uptake in plants. Those that directly benefit the plant's metabolism may have considerable potential for use as biofertilisers. The most important groups in this category are symbiotic nitrogen-fixing micro-organisms and mycorrhizal fungi. The use of microbial biofertilisers was discovered by scientists approximately a century ago and has been commercialised for the past 50 years. Despite this relatively long history, the global market for this technology has been, and still is, insignificant compared to the market for chemical fertilisers. In addition, compost, which results from the activity of waste-degrading micro-organisms, is an important biofertiliser which has been used worldwide by farmers for centuries.

## Biological nitrogen fixation

Harvesting a plant removes nitrogen from the soil. The natural processes for replenishing nitrogen are too slow to sustain productivity at the level demanded by intensive agriculture. The deficit is often made up by adding chemical fertilisers. However, some micro-organisms are capable of fixing atmospheric nitrogen and so of helping to maintain or restore soil fertility. Among those of proven interest to agriculture are the soil bacteria *Rhizobium* and *Bradyrhizobium*, and to a lesser extent *Azotobacter*, *Azospirillum* and the cyanobacteria (blue-green algae) (Giller et al., 1994; Postgate, 1990; Sprent and Sprent, 1990).

### *Rhizobia*

These bacteria enter into a symbiotic relationship with leguminous crops such as bean, groundnut, soybean, pea, lupine and clover. Wherever a host plant and its corresponding rhizobia are present, biological nitrogen fixation (BNF) will occur naturally. Effective BNF enables the use of costly and eventually polluting nitrogen fertilisers to be avoided. If the process is not effective enough, either the seed or the soil can be treated with an inoculant.

The appropriate use of a rhizobial inoculant depends on three main criteria:

- nitrogen should be the major yield-limiting nutrient,
- the inoculum should experience little or no competition from the native rhizobial population in the soil,
- the inoculum should be able to survive in the soil and infect the host.

The low yield response of leguminous crops to inoculation is often caused by the fact that nitrogen is not the only yield-limiting nutrient. Many tropical soils, particularly in marginal areas, are also deficient in phosphorus and potassium and may be highly acidic. On these soils, yield increases can only be obtained when phosphorus, potassium and lime (chalk) are also added to the soil.

Rhizobial inoculation has been most successful with soybean. Many soils do not naturally contain the required rhizobial strains for this crop, which is a recently introduced one in many countries. As a rule, about 300 g of inoculant can be mixed with 65 kg of seed, an amount sufficient for soybean cultivation on one hectare of land (Hoben and Somasegaran, 1992). In this way, savings in chemical fertiliser costs estimated at 30–300 times those incurred without inoculant can be achieved (ATI, 1988; de Jaeger, 1989; Verma and Dube, n.d.).

Other legumes do not respond as spectacularly to rhizobial inoculation as does soybean, doubtless because of the presence of competing rhizobial strains in the soil. Only in the cases of reclaimed land, regularly flooded soils or after a severe drought may rhizobial inoculation lead to significant yield increases in other legume crops (van Rossum, 1994; Giller et al., 1994).

There is much scope for improving and extending existing technologies without recourse to genetic modification. Three major approaches are possible:

- agricultural practices can be improved to enhance nitrogen fixation by rhizobia,
- more efficient and competitive rhizobial strains can be selected,
- plant varieties can be selected or bred that are better adapted to symbiosis with the most effective nitrogen-fixing bacteria (Postgate, 1990).

Taking the last of these approaches first, the genetic potential for nitrogen fixation could be improved considerably in many legumes. Giller et al. (1994) point out that some breeders "have selected against nitrogen fixation in the course of improvement of crop yields by applying heavy dressings of N fertilisers to their fields". A great deal of work on the world's major food legumes, and on some feed legumes, is conducted by various centres of the Consultative Group on International Agricultural Research (CGIAR), which have global mandates for research on phaseolus beans, lentil, chickpea, cowpea, pigeonpea, soybean and groundnut. Most research on the genetic potential for nitrogen fixation focuses on the grain legumes, because of their economic importance. Research on pasture, shrub and tree legumes is extremely limited, although there is a potential for economic impact, particularly in the case of feed and fodder crops.

The poorly understood ecology of inoculants in the soil is still a major obstacle to increasing BNF through the second approach, the use of more efficient microbial strains. Strategic research on the biochemistry of BNF, including rhizobium-legume interaction and the competition between strains, is conducted by several research groups, mainly in industrial countries but also at international level and in some developing countries. Much of the applied research on rhizobial BNF is carried out under the auspices of the Microbiological Resources Centres (MIRCENs) project (2) of the United Nations (Clark and Juma, 1991). Worth mentioning is the international multidisciplinary programme known as NifTAL (Nitrogen Fixation for Tropical Agricultural Legumes), which is coordinated by one of the MIRCENs, the University of Hawaii. Researchers have attempted to modify rhizobia strains through genetic modification to counteract ammonia repression, increase the efficiency of nitrogen fixation and introduce other traits, such as streptomycin resistance and drought tolerance (Das, 1991).

Research on nitrogen fixation in most industrial countries is limited in scope and is restricted entirely to the public sector. Such research is unattractive to the private sector because the amounts of nitrogen fixed by legumes are not likely to be sufficient to sustain the production levels of intensive agriculture. Even if they were, BNF technologies would have to compete with relatively low-priced mineral fertilisers.

Non-leguminous crops, such as the highly important cereal crops, are as yet incapable of symbiosis with rhizobia, lacking the required metabolic pathways and genetic constitution. Research is under way to apply genetic modification to this problem, but progress has so far been slow, largely because the metabolic pathways are highly complex, involving numerous genes. Currently, the best option for enabling cereals to use biologically fixed nitrogen is to rotate them with legumes (Giller et al., 1994; Lynch and Hobbie, 1988). Another option is to intercrop a cereal with a legume, although in this case the nitrogen contribution of the legume is likely to be lower.

Many developing countries, including Thailand, India, Bangladesh, Indonesia, Kenya, Zimbabwe and Brazil, have long had facilities for producing and conducting research on rhizobial inoculants. Inoculants for soybean account for a large proportion of total inoculant production. In some countries, inoculants for other grain legumes and for some tree and pasture legumes are also produced. Most inoculants are packed in polythene bags weighing 100 g, which are sold for around US$ 1.00. For a small-scale production facility producing 100 000 bags per year, the initial investment costs in buildings and equipment are relatively low, about US$ 50 000–100 000. Costs for a large-scale plant amount to an estimated US$ 2 million. Major production problems include: maintaining sterile production conditions and pure seed cultures; assessing the quality of cultures during and after manufacture; coping with the seasonal character of demand, which results in long periods during which the equipment is not used; and ensuring adequate cold storage and transport of the inoculant. (If inoculants are exposed to high temperatures or direct sunlight at any stage between processing and planting, they may become completely useless.)

The degree to which externally applied inoculants are an accepted farm practice in developing countries varies considerably. In all cases, however, the use of inoculants is limited mainly to larger scale producers (Eaglesham, 1989). Inoculants have had little effect on legume production where yields are poorest, at the small-scale farm level in the marginal areas, the main reasons being that nitrogen is usually not the only or major yield-limiting factor in such areas and that inoculant strains are often not well adapted to conditions there.

*Cyanobacteria*
These organisms, otherwise known as blue-green algae (BGA), can associate with almost all groups of plants, although most grow and fix nitrogen independently of their host (Sprent and Sprent, 1990). Many associations between cyanobacteria and their hosts are, how-

ever, of little benefit to the host, and are therefore irrelevant from an agricultural point of view. Cyanobacteria require a high level of soil moisture, even flooding, if they are to flourish. In terrestrial systems there is little evidence of agriculturally significant nitrogen fixation by cyanobacteria. Of interest to agriculture, particularly to rice production, are *Anabeana azolla* associations, and a few of the many genera that fix nitrogen independently.

The symbiotic cyanobacterium *Anabeana azolla* associates with the water-fern azolla, triggering BNF. Azolla is one of the few ferns adapted for aquatic living and is native to Asia, Africa and South America, where it can be found in fresh-water environments such as ponds, canals, drainage ditches and, of course, rice paddies. After initial growth in nurseries, azolla can be intercropped in paddy fields. Under ideal conditions, it can add more than 10 kg of N/ha/day, although under field conditions about one to two kilograms of N/ha/day is a more realistic expectation (Sprent and Sprent, 1990). Azolla also has high organic matter and protein contents. Once dead, it decomposes rapidly in the soil, releasing about 50% of its nitrogen compounds in three weeks. In this way azolla can replace up to 30 kg of urea per hectare. Apart from its use as green manure, azolla has long been used as fodder for poultry and livestock. It is estimated that azolla is currently used in only two per cent of the world's rice paddies, mainly in China, Vietnam, Egypt, India and the Philippines, and that its use is diminishing, the main reasons being its high labour requirements, sensitivity to certain insect pests, vulnerability to chemical pesticides and high phosphorus requirements, combined with the increased availability and use of chemical fertilisers.

A few other genera of cyanobacteria, such as *Nostoc, Aulosira* and *Calothrix*, can also be used as biofertilisers in the cultivation of paddy rice, supplementing the use of mineral fertiliser. Yield improvements of 5–25% have been achieved, even when 100 kg of N/ha has been added as fertiliser (Sprent and Sprent, 1990). This suggests that cyanobacteria may have other benefits besides adding nitrogen. These may include the detoxification of soil by the removal of sulphide and ferrous iron, and/or the excretion of ascorbic acid, both of which may stimulate rice growth.

Use of a BGA inoculant, as opposed to the algae already present, is still not widespread, but it is practised in some parts of India (Sprent and Sprent, 1990). BGA is produced by farmers, using shallow trays or pits filled with soil and water and covered with a thick polythene sheet. The sun-dried algal material can be stored in bags until application, without loss in viability. Generally, farmers need to apply inoculant for three to four consecutive cropping seasons to build up an effective population of algae. Further inoculation is only necessary if unfavourable ecological conditions occur (Venkatamaran and Shanmugasundaram, 1992). Formal research on the selection and improvement of inoculant strains is only just beginning.

*Azotobacter and Azospirillum*

In the rhizophere of cereals (such as maize, sorghum and millets) in particular, but also of some other crops such as sugar cane and sunflower, members of the *Azotobacter* and *Azospirillum* genera of bacteria are found. Their beneficial effect on plant growth is only partly due to their ability to fix atmospheric nitrogen. More important factors are their ability to stimulate root development and the production of plant growth substances, and to enhance the uptake of other nutrients. These properties are also apparent in other non-nitrogen fixing rhizosphere bacteria, such as certain *Pseudomonas* species. Under certain conditions, adding inoculants of *Azotobacter* or *Azospirillum* to the soil can increase crop yields. In several developing countries, such as India, inoculants of these species are produced and applied successfully on a small scale (Verma and Dube). Understanding of the physiological mechanisms at work in these associations is limited and few research groups are working on these bacteria. Research in industrial countries is negligible, apparently because *Azospirillum* does not "work" in soils in temperate climates.

## Mycorrhizal associations

Mycorrhizal associations are a form of symbiosis between certain fungi, particularly vesicular-arbuscular (VA) mycorrhizae, and the roots of higher plants. VA mycorrhizal fungi are non-specific and can infect a very wide range of host plants. It is assumed that more than 90% of all vascular plants can form mycorrhizal associations. In many circumstances, mycorrhizal infection can greatly increase the uptake of nutrients, particularly phosphorus and nitrogen, from deficient soils. In addition, it may mobilise trace elements such as copper, zinc and, possibly, iron (Stribley, 1989).

Although mycorrhizal fungi are generally widespread in soils, artificial inoculation may improve host response where the concentration of natural populations is locally suboptimal (which is almost always the case where new land is taken into cultivation), or when the natural population is inferior in its physiological effects on the host. Large-scale production of pathogen-free mycorrhizal inoculants in aseptic culture is not yet possible. Commercial inoculants have been produced and marketed to a very limited extent in the USA.

For reasons that remain unknown, the use of artificial mycorrhizal inoculants has not so far proved very effective in increasing crop yields. The understanding of mycorrhizal associations is extremely limited: little information is available on the genetic variability and distribution of natural populations of VA mycorrhizal fungi, or on environmental effects on the natural selection of particular genotypes. Artificial inoculation may be unnecessary if an effective population of native mycorrhizal fungi in the soil can be built up through improved management practices.

Relatively little research is conducted on mycorrhizal associations worldwide. A considerable increase in knowledge of the physiology of these organisms will be required before science-based biotechnology will be able to contribute effectively.

## Composting

Composting is the aerobic solid-state degradation of organic matter by a mixed population of micro-organisms in a warm, moist and aerated environment. Composting of animal excreta mixed with drier material, such as domestic garbage or crop residues, has been practised by many farmers all over the world for centuries (Lynch and Hobbie, 1988). In the developed world, a compost heap is often infected with a starter culture, which usually contains *Azotobacter*. In many developing countries, however, starter cultures are not available, in which case composting may be a rather slow process. One exception is India, where phosphate-solubilising micro-organisms (PSM), such as members of the genera *Bacillus* and *Pseudomonas* , are produced and marketed on a limited scale (Verma and Dube, n.d.). PSM-inoculant is mixed with the compost materials at the time of first turning (after one and a half months).

A simple compost heap might consist of wooden supports with shelves or netting to allow ventilation. Large-scale composting processes are, by constrast, much more complicated. They require several parameters to be controlled, including acidity, air flow, moisture content and the initial carbon-nitrogen ratio. Several problems arise in making and using compost:

- The collection of the necessary organic matter is labour-intensive.
- The composting process requires reasonable amounts of moisture. In many (semi-) arid areas, water availability is low, and the process is slowed down to such an extent that many farmers do not find it worthwhile.
- If the temperature does not reach at least 60°C for 24 hours, pathogens and weeds may develop in the compost.
- If not fully matured, the compost may make its own demand for nutrients as breakdown continues in the soil.
- Compared with mineral fertilisers, compost is a dilute source of nitrogen, phosphorus and potassium; it may well be impossible to prepare enough compost to meet all the needs of the crop for these major nutrients (Dalzell et al., 1987).

Relatively little formal research on composting takes place in either the developed or the developing world. Most advances occur at the farm level. For example, a modified anaerobic composting system, in which aerobic solid state fermentation is combined with an anaerobic decomposition process, has been developed by a farmer (see Kuruvinakunnel, 1994; Nazareth, 1994).

## Biocontrol agents

Certain bacteria, fungi and viruses are natural pesticides. In addition, plants and animals may produce substances which are instrumental in the control of various pests and diseases, such as certain biocidal peptides, insect pheromones and plant extracts. Pheromones and plant extracts will be discussed in this section because biotechnology may play a role in their production in the future, although it does not do so at present.

The basic principles of biological control were established about 60 years ago, yet this technology has remained overshadowed by chemical control. The use of agrochemicals appears to have become indispensable in many areas where modern intensive agriculture is practised. However, agrochemicals may harm the environment and pose health risks to both the consumer and the farmer (DiTomaso et al., 1993). Another major drawback is that target organisms may develop resistance. These adverse effects have led to a growing interest in biocontrol agents, on the part not only of alternative farming and environmental groups but also of farmers involved in intensive agriculture and of policy makers eager to save scarce foreign exchange.

## Bio-insecticides

Bio-insecticides can be divided into microbial insecticides (bacteria, fungi and viruses), pheromones and plant extracts.

### Microbial insecticides

Bacteria form the most widely used microbial insecticides, mainly because of the high efficacy of certain strains of the infectious spore-forming bacterium *Bacillus thuringiensis* (Bt). Bt accounts for over 90% of all microbial insecticides. It produces crystalline proteins that together are toxic to many insect species, mainly caterpillars, beetles, mosquitoes and flies (Coleoptera, Lepidoptera and Diptera) (Barnett et al., 1993; Ely, 1993; van Frankenhuyzen, 1993; Macdonald, 1989). The first strain of Bt was isolated in 1911 and Bt strains have been commercially produced since the 1960s.

Bt-based insecticides have several disadvantages: the toxin has a short shelf-life; it deteriorates rapidly in the field; and it is quite expensive. Research has focused on improving the persistence of the Bt toxins in the field. Bt toxin genes have been successfully transferred to the bacterium *Pseudomonas fluorescens*. The fermented *Pseudomonas* cells are gradually killed by heat and chemical reaction, resulting in slower breakdown and longer-term release of the toxin. A further and major disadvantage is that resistance to Bt proteins has recently developed in several species, including the Indian meal moth, the diamond back moth and the Colorado potato beetle, following high-dose spraying over a period of several years (Commandeur and Komen, 1992). This problem has led a consortium of agrochemical and biotechnology companies to form the Bt Management Working Group to devise more effective regimes for the use of Bt products. One way of overcoming resistance may be to use two agents simultaneously, for example two Bt products or one Bt product and one alternative chemical or biological product (for example proteinase inhibitors).

Most of the substantial and increasing amount of research on Bt, both strategic and applied, is done in the USA and Europe and is directed to crops and pests of importance to these regions. Several developing countries also have a limited research capability in this area, notably Kenya, Egypt and China. Research in these countries is usually adaptive in nature, focusing on the selection and mass production of effective strains.

Options other than spraying Bt include the incorporation of genes coding for Bt proteins in host plants. Research in this area was conducted for tobacco and tomato in the mid-1980s, and recent work has extended the list to other crops, including maize, rice, soybean, cotton, potato and poplar (Koziel, 1993). The advantages of incorporating insect resistance in this way are that:

- the toxin is present at the onset of infestation,
- all tissues are protected, even when the insect is located within the host plant,
- protection is independent of the weather.

One disadvantage is that each new crop variety has to be engineered separately.

The fungi associated with insects can also be used as control agents, but they have not been widely commercialised (Macdonald, 1989). Many fungi are unique among insect pathogens, as they do not need to be ingested, being able to penetrate directly through the arthropod cuticle. A few strains of pathogenic fungi are already applied to the control of insects. Examples are *Beauveria* spp, which have about 500 host species, mainly Lepidoptera and Coleoptera, and *Metharizium anisoplae*, which has more than 200 known hosts among the Coleoptera, Lepidoptera, Orthoptera and Hemiptera (Heiny and Templeton, 1993). Spores of *Metharizium anisoplae* are sprayed to control certain insect pests of rice and sugar cane. This organism forms the basis of a commercial bio-insecticide (BIO 1020) produced by Bayer. A similar product is manufactured locally in developing countries, for example in northeast Brazil, by small companies and farmers' cooperatives.

Aside from these few examples, fungal pesticides are little used in either developed or developing countries. The main factors constraining further use are the relatively high costs of mass production, the instability and short shelf-life of fungi-based products, their poor persistence in the field, their requirements for very moist conditions and their slow rates of killing.

The use of viruses in insect control is constrained for similar reasons. Recent advances in genetic modification have led to the development of insect viruses that express neurotoxins from the mite *Pyemotes tritici* and the scorpion *Adroctonis australis*. This new generation of viruses slows down the feeding rates of insects, but the results are not yet impressive enough to enable the products based on them to compete with chemical insecticides or with Bt products. There are also biosafety risks associated with these products, namely the possibility of mutation or transfer of the toxin gene to other organisms.

### Pheromones

These are signal compounds used by insects to communicate. They allow mates to find each other, locate food, alert against danger, and so on. In several cases pheromones have proved extremely useful in pest control by:

- luring insects to their deaths, when mixed with insecticide,
- disrupting mating (this is achieved by distributing the attractant pheromone widely in the insect's environment so that the sexes are unable to locate one another),
- attracting insects into monitoring traps, enabling the farmer to determine whether spraying is needed (McDonald, 1991; Whitten and Oakeshott, 1991).

The most efficient way of controlling insect populations is the lure-and-kill technique. Mating disruption only works with isolated populations, which are rare.

The only widespread application of pheromones is the use of pheromone-based monitoring systems. These are already available commercially from a number of companies in Europe and the USA. All the pheromones currently marketed are made by chemical synthesis, but biotechnology (enzyme technology) is of potential interest since it could bring an efficiency gain. Pheromones consist of two stereo-isomeric variants (mirror-image molecules with the same chemical formula), only one of which has the desired biochemical effect. Using enzymes will allow the effective form only to be produced, whereas in chemical synthesis both variants are produced.

### Plant extracts

Crude plant extracts have also proved effective in pest control. Examples include the use of curcuma powder against the pulse beetle (Ahmed and Ahamad, 1992), mustard oil against the rice weevil (Chandler et al., 1991) and rotenone, produced by the small shrub *Derris elliptica,* against a variety of insects and nematodes (van Latum and Gerrits, 1991). The most widely used plant compounds are pyrethrins, produced by *Chrysanthemum cinerariifolium* (van Latum and Gerrits, 1991), and neem oil, produced by the neem tree, *Azadirachta indica* (Grace and Yates, 1992). The former is used against a wide variety of insects, including aphids, mosquitoes, flies, cockroaches and grasshoppers, while the latter is effective against sucking insects, butterflies and aphids.

Pyrethrum is more expensive and less effective than neem oil, but it is better established commercially. Several developing countries, among them Kenya, Tanzania and Ecuador, now grow *Chrysan-*

*themum cinerariifolium* for export. Kenya's export earnings from pyrethrum are around US$ 20 million annually. Production levels are, however, not sufficient to meet a highly fluctuating global market demand. The neem tree is native to India and Burma, but has been introduced to many other countries in the tropics (National Research Council, 1992). Neem products have been used for centuries by farmers on the Indian subcontinent to protect their crops from insect attacks. Simple extracts of ground leaves or seeds can be very effective, and production is uncomplicated.

Research on plant extracts in developing countries has focused on increasing natural production and improving extraction rates of the toxic compounds described above. Efforts in the developed world concentrate on the use of gene transfer to produce the compounds by other means. To expand production, the genes coding for the toxic compounds have been transferred to bacteria or introduced directly into food crops. In the case of pyrethrin, several research groups have attempted to produce these compounds *in vitro* using plant cell tissue and organ cultures (Jovetic, 1994). So far, this research has not led to commercially interesting results, since production levels in these cultures are still very low compared to those in whole plants.

## Biofungicides, biobactericides and bioherbicides

The development of microbial control for weeds and diseases has lagged behind the control of insect pests in both research and development. Few commercially successful applications are available. Limited use is made of a number of micro-organisms applied as herbicides. The oldest is the fungus *Collectrotrichum gloeosporoides*. This is marketed under the name Collego and is pathogenic for northern jointvetch, a weed of rice and soybean.

Recently, a microbial fungicide called Mycostop has become available (Anonymous, 1993). This product contains the bacterium *Streptomyces griseoviridis*, and is applied as dry powder (for seed dressing) or in aqueous suspension (for soil spraying, drenching or root dipping of cuttings) to control *Fusarium* diseases on flower crops such as carnation, gerbera and cyclamen, and on the vegetable cucumber. Toxicological tests have shown Mycostop to be a safe product. Unfortunately, it is just as expensive as the agro-chemicals it could replace.

Microbial inoculants are also used against bacteria. A well known example is the fungus *Trichoderma*, used to prevent Dutch elm disease (Macdonald, 1989; Campbell, 1989; Baker, 1989). Other promising inoculants are the plant-growth-promoting rhizobia (PGPR), fluorescent pseudomonads and strains of *Bacillus subtilis,* which appear to be effective competitors of minor root pathogens. They may act by reducing the availability of iron for deleterious rhizobacteria or fungi, or by producing toxic compounds (Campbell, 1989; Kloepper et al., 1989). Box 7.3 provides an example of research in this area.

**Box 7.3**
**Biocontrol of cassava root rots**

Cassava is a crucial food security crop in some of the poorest parts of the world. As pressure on land increases, it is often planted in succession in the same field, a practice which increases the incidence of cassava root rots. These rots directly attack the starchy storage roots underground, where the problem remains hidden until harvest. Because root rots are commonly caused by a complex of micro-organisms, including fungi (*Fusarium, Phytophthora, Diplodia* and *Pythium*) and bacteria (*Pseudomonas and Erwinia*), it is difficult to identify specific host-pathogen interactions on which to focus resistance breeding. Chemical treatment of planting materials offers partial control, but is uneconomical for most small farmers.

In search of a solution, scientists at the Centro Internacional de Agricultura Tropical (CIAT), based in Cali, Colombia, first looked at farmers' fields where cassava had been successfully grown for unusually long periods of time. They argued that these fields would be likely sources of effective natural enemies. Soil from fields in different environments in Latin America was sampled. In addition, soil was sampled whenever healthy roots were harvested in heavily infected cassava fields.

In laboratory and greenhouse experiments, several strains of bacteria (*Pseudomonas putida, P. fluorescens*) and fungi (*Trichoderma harzianum*) found in the soil samples effectively reduced the population of root rot organisms. Since *Trichoderma* species are easier to handle and propagate than *Pseudomonas*, they were selected as the most promising candidates for inoculum production.

Results of initial field trials indicate that the micro-organisms in the inoculum have problems competing with well adapted micro-organisms already in the soil, a problem also encountered in biological nitrogen fixation. As expected, success depends on many environmental factors. Researchers believe that trials in up to 15 or 20 different sites are necessary to evaluate the true potential of a given strain of micro-organism.

Source: A-M Thro, coordinator of the Cassava Biotechnology Network, CIAT (personal communication).

## Advantages and disadvantages

Microbial inoculants have several advantages over chemicals in the control of pests and diseases (Table 7.1).

Both McDonald (1991) and Reeves (1991) estimate that biological control agents, particularly bio-insecticides, could reduce global use of agrochemicals by as much as 80%. The exploitation of micro-organisms certainly looks like a positive development in biological control. However, the fact that microbial inoculants still account for only a small percentage (<5%) of the market suggests that they do have certain disadvantages:

● The use of bio-insecticides needs precise timing, since they generally attack at only one stage in the insect's life cycle. Farmers who use these products similarly to agrochemicals may well not achieve the desired effects.
● The shelf-life of microbial inoculants is shorter than that of chemical sprays, making product stocks a concern to both trader and farmer (Netzer, 1987). Since microbial inoculants are alive, they are more sensitive to environmental changes than chemicals.
● To enhance their efficacy, inoculants must often be combined with chemicals.
● The cost of most biocontrol agents (except for plant extracts) is not currently lower than that of most chemicals.

TABLE 7.1   Comparison of chemicals and microbial biocontrol agents

|  | *Chemicals* | *Microbial biocontrol agents* |
| --- | --- | --- |
| *Cost/benefits* | | |
| R&D | US$ 20m | US$ 0.8–1.6m |
| Market size required for profit | US$ 40m/year to recoup development costs, therefore limited to major crops | Markets under US$ 1.6m may be profitable due to low development costs |
| Toxicological testing | US$ 10m | US$ 0.5m |
| Patentability | Well established | Still developing |
| Lead time | 6–7 years | 3–5 years |
| *Discovery* | Screen 15 000 compounds to identify one product | Rational selection for specific target pests |
| *Efficacy* | | |
| Kill | ca. 100% | Usually 90–95% |
| Speed of kill | Rapid | Can be slow |
| Spectrum of activity | Generally broad | Generally narrow |
| Resistance | Often develops | Only few known cases |
| Type of action | Can be both preventive and curative | Generally only curative |
| *Safety* | | |
| Operator safety | Chemicals can be hazardous | Low operator risk |
| Environmental impact | Many examples, e.g. accumulation in food chains | Few examples with use of indigenous micro-organisms |
| Residues | Interval before harvest often required | Crop can usually be harvested immediately after application |

Source:  Whitton and Roger (1989) .

- The development of new biocontrol agents is a long process requiring significant investment. Most of the microbial inoculants currently on the market have been worked on, with varying intensity, for about 20 years (Campbell, 1989).
- Companies with the financial and technical clout to develop microbial inoculants usually also have synthetic products to sell. This lowers their enthusiasm for biocontrol agents.

The use of inexpensive and locally available organic substrates (for example fermented cassava, maize or cowpea) combined with cheap labour makes the production of microbial biocontrol agents a practical proposition in several developing countries. China, Thailand, India, Egypt, Nigeria, Brazil and Mexico are among the countries that have already undertaken production of

Bt (Salama and Morris, 1993). The costs involved in obtaining registration and approval for microbial pesticides are relatively low compared to those for chemical pesticides.

Plant-derived biocontrol agents are relatively cheap and are easier to produce. The neem tree is the most obvious example of an option well suited to the needs of small-scale farmers. The use of neem could be promoted by giving each farm or homestead its own neem tree(s) for use as timber, firewood, charcoal, medicine, insecticide, nitrification inhibitor or feed. It remains to be seen whether science-based biotechnology can play a more significant role in the production of plant-derived biocontrol agents in the future than it has done so far.

## Diagnostics

Antibodies have long been used both in research and in plant and animal health care as tools for detecting the presence of specific organisms or proteins, in particular those of a pathogenic nature. In the 1980s a new category of immuno-assays was developed, which included the use of monoclonal antibodies (see above). This was followed by the development of a new approach to diagnosis using DNA probes.

### Immuno-assays

The principles of monoclonal antibody (MAb) technology have already been outlined. The main use of MAbs in crop production is in the diagnosis of viral and bacterial diseases. Easy-to-use diagnostic kits could be produced in developing countries. Where labour is cheap, the unit price could be as low as US$ 12–15 per 100-test kits. Yet none of the developing countries have so far undertaken commercial production.

The appropriateness of diagnostic tests depends greatly on their price, together with the economic importance of the disease and the options open to farmers for actually controlling a disease once it has been diagnosed. Many small-scale farmers are more constrained in the latter respect than their large-scale colleagues. The impact of MAb tests is therefore likely to be greatest on large-scale agriculture.

### DNA probes

DNA probes are short fragments of DNA (often less than 500 base pairs) that attach themselves to a piece of DNA or RNA with a sequence complementary to their own one. Probes can be tagged with a radioactive, fluorescent or enzyme label, and in this way they can be used to detect the presence of specific DNA sequences in organisms. These sequences may reveal the presence of infectious agents, or demonstrate the genetic basis of certain traits. Tests for infectious diseases in animals, plants and humans are relatively easy to produce using DNA probes, and have therefore found considerable commercial application.

The main advantages of DNA probes in diagnostics are their high specificity, sensitivity and speed. They often reveal pathogens that would not have been detected by immuno-assay. Continuing disadvantages are the relatively high level of skills required to develop and use DNA probes, necessitating considerable investment in training, equipment and quality control. Currently, DNA probe-based tests are quite expensive, each costing about US$ 5.00. However, costs are likely to reduce in the near future, encouraging application in developing countries. If such application is to be widespread, tests will have to prove both stable and sensitive. They will also have to use colourimetric rather than radio-isotope labels: being radio-active substances, isotopes pose a health risk and are therefore difficult and expensive to handle. Only a few DNA probe-based tests fulfil these requirements as yet, but as the technology evolves their number will undoubtedly rise.

Several technologically more advanced developing countries (e.g. Thailand and India) are conducting research on the diagnostic use of DNA probes. For the time being, DNA probes are too expensive for direct use in small-scale agriculture. They are, however, being used by institutes with a mandate for improving crop production among small-scale farmers. For example, polymerase chain reaction (PCR) is being used by the Centro Internacional de Agricultura Tropical (CIAT) to detect the presence of viruses or families of viruses in *Brachiaria* pastures, beans and other crops.

# Animal nutrition and health

The productivity of an animal depends on its genetic make-up, the feed resources available to it and the use it makes of them, and the biotic and abiotic stresses to which it is subject, especially diseases. Science-based biotechnology allows progress in all these areas.

## Animal nutrition

Ruminants, poultry and pigs make a huge contribution to the well-being of people in developing countries. They are usually fed on poor-quality forages, crop residues and pastures produced on relatively infertile land, and thus do not compete with human beings for food staples. Advances in animal nutrition could, if widely applied, bring about considerable gains in livestock productivity, which is generally low at present.

Science-based biotechnology interventions are available at three different points in the livestock production system: feed before consumption, the digestion process within the animal, and the metabolism of the animal.

## Feed quality

One fairly widespread application of science-based biotechnology is the addition of micro-organisms and/or their products to animal feeds to improve their preservation and digestibility. Of special interest, but still undergoing research, is the use of enzymes in feed. These proteins can be added to degrade compounds which cannot be adequately broken down by the animal, thereby increasing the nutritive value of the feed and/or minimising the amount of waste produced. The most promising enzymes are those capable of:

- degrading structural carbohydrates, such as cellulose and lignin,
- degrading or structurally modifying proteins in such a way that these loose their anti-nutritional properties,
- liberating phosphorus from complex compounds.

An example of the last category is the enzyme phytase, which is already produced commercially (see Box 7.4).

The primary constraint to ruminant production using fibrous feed is the low efficiency of feed utilisation. Preliminary studies indicate that pre-digestion with degrading enzymes makes cell walls available as a source of energy. Moreover, digestion of the cell contents is also improved, providing significant extra amounts of protein and minerals. The digestibility of feeds such as grass silage is limited by the presence of lignin, which acts as a physical barrier to microbial utilisation of cellulose and hemicellulose by rumen micro-organisms. Lignin can be degraded in the solid state by enzymes directly or by bacterial inoculation (Mudgett, 1986). If fungi could be used as well, this would not only improve digestibility but also increase the protein content of the feed through the production of fungal protein. However, these applications are largely still at the research stage. An important constraint is that enzymes need to be found or developed

---

*Box 7.4*
*The case of phytase*

Phosphate is an important nutrient for monogastric species such as pigs and poultry, but it is generally present in their feeds in the form of phytic acid, a compound which these non-ruminant species cannot degrade. Phosphate from phytic acid is, therefore, mainly excreted and is a major component of manure. In many developed countries this has led to the pollution of soil and water resources. If uptake by the animal could be increased, this problem could be substantially reduced.

The private sector in the developed world has conducted extensive research to reduce the excretion of phosphate by monogastric animals. Phytase, produced and marketed by Gist-Brocades, a large Dutch biotechnology company, is an enzyme capable of converting phytic acid into inorganic phosphate and inositol in pigs and poultry. Recently, the company has improved the expression of the gene coding for phytase in *Aspergillus*, thereby considerably reducing production costs.

This technology is widely available in the developed world, but is not yet manufactured or used in developing countries.

Sources: Beudeker (1990); Bos (1990).

which result in better penetration of the cell wall. The process is currently too slow to be economically viable (Rai et al., 1988).

Most of this type of research is conducted in the developing world, particularly in Asian countries such as Sri Lanka, India and Thailand. None of it has yet led to applicable results.

There are some general problems with the application of enzymes in animal feed. The feed is often too dry or is stored at the wrong temperature before consumption for the enzymes to become active once the feed enters the animal. Inside the animal's digestive tract, temperature and moisture conditions are likely to be appropriate, but the animal's own digestive enzymes may attack exogenous enzymes and render these inactive. The low pH in the stomach may also adversely affect the action of introduced enzymes.

## Digestion

Research to improve digestion aims to influence the action of rumen microflora through the addition of micro-organisms, antibiotics and other active substances that stimulate the production of saliva and other digestive fluids. None of these innovations has yet reached the stage of widespread application. A major constraint is the difficulty of manipulating the microflora of the digestive tract, which is a very complex system not well understood by researchers. The relevance of these innovations for small-scale farmers will remain dubious for many years.

## Metabolism

During the past 20 years, genetic modification techniques have been applied in the laboratory to the production of active substances that influence the metabolism of several domestic animal species. An example is the production of the hormone bovine somatotropin (BST), a protein which is naturally produced by the pituitary gland of cows (MacKenzie, 1988). BST enhances growth, increases lactation and influences the balance of carbohydrate and fat metabolism. Injected into cows, it increases milk production by 10–20%, partly by boosting the cow's appetite, partly by diverting a larger proportion of her food from ordinary metabolism to the production of milk. In the late 1970s, scientists inserted the gene for BST into bacterial cells (*Bacillus subtilis*), which enabled the large-scale production of BST by the multinational Monsanto (MacKenzie, 1988). Other examples are chicken growth hormone and porcine somatotropin (PST), which alter the protein/lipid ratio in the animal.

The use of BST and similar products by farmers is still illegal in the European Union, where there is strong opposition from producers and consumers, owing partly to the region's current over-production of milk. In a few other countries, notably the USA, BST has recently been approved. The use of BST in developing countries is still very limited and its relevance to the small-scale dairy farmer is doubtful. The effectiveness of BST and similar products depends on first-rate management, high-quality feeding regimes and good animal health, all of which require high levels of finance. These conditions are seldom met in small-scale agriculture, especially that of resource-poor farmers in marginal areas.

## Animal health

We will restrict our discussion of animal health mainly to the development of drugs and vaccines. The diagnosis of animal diseases will be discussed only briefly, since the techniques used largely overlap with those already covered in the section on plant diseases.

### Drugs

Several micro-organisms produce secondary metabolites (compounds which are not essential for survival and reproduction, and whose synthesis occurs after growth ceases) which are important drugs in veterinary health care. The most important group of secondary metabolites are the antibiotics, which inhibit or kill other micro-organisms. To be effective, antibiotic molecules must interfere with a fundamental metabolic process of the pathogenic micro-organisms without (ideally) affecting the animal host. The main antibiotic producers are fungi, actinomycetes (bacteria which grow in mycelial form) and eubacterial species (Antebi and Fishlock, 1985). The first antibiotic, penicillin, was discovered in 1929 and is produced by the obligate aerobic mould *Penicillium notatum*. Other appropriate micro-organisms have since been obtained by screening soil, plant or water samples for antibiotic activity. Fewer than one out of 10 000 screened strains provides a useful antibiotic (Antebi and Fishlock, 1985).

When a suitable strain has been selected, it is cultivated in a fermentor, and the antibiotic is purified. If the biological quality, safety and efficacy of the antibiotic are acceptable, strain improvement (either through conventional mutation techniques or through genetic modification) and fermentation development are undertaken. Much of today's antibiotic production involves semi-synthetic processes in which microbes are used to produce a precursor compound which subsequently undergoes chemical modification. The major problem in the use of antibiotics is that target organisms frequently develop resistance.

Apart from antibiotics, there are many other secondary metabolites with useful characteristics. In veterinary health care, microbial products include coccidiostatics, anti-helminthic drugs and anabolic muscle-enhancing drugs. The range of non-antibiotic microbial products is likely to increase as sensitive and rapid assays for a range of biological activities are developed.

Production of some drugs in developing countries is technically feasible and economically attractive. Sufficient expertise in microbiology and a good quality control system are, however, essential. Local production would reduce the prices of products to small-scale users but, as in some Western countries, overuse might then become a problem.

## Vaccines

Animals are often vaccinated to prevent infectious diseases. Vaccines must result in immunity without themselves being pathogenic. It is this separation of immunogenicity from pathogenicity which is the crucial aspect of vaccine development.

Conventional vaccines can be divided into two main categories. The first consists of replicating vaccines, made from live attenuated organisms. By culturing the original pathogen in the laboratory for numerous generations under artificial conditions, as in cell culture or liquid media, the organism may lose most of its virulence while retaining immunogenicity. However, replicating vaccines may revert to full virulence. The second category contains non-replicating or inactivated organisms. Controlled inactivation by heat or chemical treatment can destroy pathogenicity while retaining immunogenicity. These vaccines may consist of only part of an organism (usually one or more proteins), in which case they are known as sub-unit vaccines.

The emergence of genetic modification has led to the development of a new range of vaccines. Generally, these vaccines are more stable and safe, and can be developed against diseases for which effective conventional vaccines are not available. Novel biotechnological approaches offer opportunities to improve the control of several serious human diseases, such as malaria and hepatitis, and a number of animal diseases, including Aujeszky's disease, Newcastle disease, theileriosis and rinderpest.

### Replicating vaccines

Genetic modification is used to make two types of replicating vaccines, deletion mutant vaccines and carrier vaccines. In deletion mutant vaccines, the genes coding for the virulence factors (the components of the organisms responsible for their pathological characteristics) are removed from the genome of the pathogenic organism. The mutated live organism is non-virulent, but still capable of eliciting an immune response. Since the viruses from deletion mutant vaccines can be distinguished in the field from wild-type viruses, these are the vaccines of choice in disease eradication programmes. In carrier vaccines, genes coding for immunogenic antigens of pathogenic organisms are introduced into a vector, consisting of a known non-pathogenic virus or bacterium. The bovine vaccinia virus, the capripox virus and *Salmonella* bacteria are regarded as the most promising vectors. The transformed vector expresses the foreign genetic information coding for the desired antigen. Using this method, it is possible to insert more than one foreign gene to produce a multivalent vaccine.

*Non-replicating vaccines*
Of the many antigens which a pathogen contains, only a few are responsible for inducing an immune response which leads to resistance to the pathogenic organism. Identification and characterisation of these antigens is the first step in the development of a well-defined subunit vaccine. Once an antigen has been identified, it may be partly purified from the organism and used for vaccine production. Alternatively, the researcher can track down the genes that code for part or all of it. This DNA is then isolated and characterised, before being introduced into a host genome, which is usually a bacterium, yeast or mammalian cell. If the transfer is successful, the recombinant organism will produce the desired protein or peptide in large quantitites (Muscoplat, 1989).

The first vaccine produced using science-based biotechnology was a sub-unit vaccine against neonatal diarrhoea in piglets and calves (enterotoxigenic *Escherichia coli*) developed in 1982. A more recent example is a vaccine against feline leukaemia. A disadvantage of all non-replicating vaccines is that, because they are inert and cannot be amplified in the body, they must be administered with adjuvants, compounds which boost their effects (Tartaglia and Paoletti, 1988). However, the adjuvants currently used sometimes cause toxic side-effects. A better understanding of adjuvant activity is needed before further progress can be made in this area.

Examples of deletion mutant vaccines are those against porcine Aujeszky's disease, classical swine fever virus and infectious bovine rhinotracheitis caused by bovine herpes virus type 1. The advantage of a deletion mutant vaccine is the ability to distinguish between vaccination and infection by a wild-type virus. This is done by immuno-detection of the antigen encoded by the wild-type virus but not by the vaccine strain. Distinguishing the two is essential when monitoring the spread of a wild-type virus in the field at the same time as vaccinating to eliminate the pathogen through an eradication programme.

Few carrier vaccines have yet become available. Vaccines containing genes encoding a rabies antigen or a rinderpest antigen have been tested on a small scale, while vaccines against Newcastle disease in poultry, based on vaccinia virus or Marek's Disease virus, have recently been developed. Carrier vaccines follow the infectious route of the carrier virus and usually produce no symptoms since only a single antigen of the pathogen is expressed. A potential problem with carrier vaccines is that the expression of the foreign gene(s) in a host cell may not be followed by proper post-translational processing (that is proteolytic processing and/or glycosylation), which may be important for an optimal immune response (Tartaglia and Paoletti, 1988). Finally, not all carriers are equally well suited to repeated use in the same animal, due to the immune response they provoke. This may limit their use or require multivalent vaccines to circumvent the problem.

To date, few biotechnology-based vaccines are commercially produced and none have yet been developed against parasitic diseases. The costs of developing vaccines through biotechnology are, at an estimated US$ 5–10 million per vaccine, more than ten times higher than those of a conventional vaccine. The equipment and materials needed are expensive and the personnel must be highly qualified. It takes at least five years to develop a candidate vaccine and another five years to turn it into a marketable product. Only a handful of developing countries, including India, Thailand, Zimbabwe and Kenya, currently conduct research in this field.

Nonetheless, vaccines which are safe, effective and heat-stable could be of great benefit to developing countries. Controlling some of the more severe and widespread animal diseases in such countries is important not merely for the domestic livestock sector but also in gaining access to export markets. Unless subsidised, particularly at the developmental stage, most of these new vaccines will, however, be too expensive for the small-scale farmers who most need them.

### Diagnostics

The contribution of monoclonal antibody technology to the development of diagnostic kits has already been discussed. Diagnostic kits can play an important role in maintaining animal health.

Several kits for major animal diseases, such as Newcastle disease and Aujeszky's disease, are already in widespread use in the developed countries. Other kits have been developed in the laboratory, but have so far been used only on a semi-commercial scale. Examples are kits for the detection of avian leukosis, infectious bronchitis in chickens, bovine viral diarrhoea and bovine respiratory syncytial viruses. Box 7.5 describes the development of kits to detect tick-borne diseases of cattle.

Though highly relevant to the needs of developing countries, diagnostic kits for animal diseases are likely to be widely used in such countries only if part of an intensive control and eradication programme.

# Germplasm improvement and conservation

A wide range of science-based biotechnologies has been developed for use in the crucial areas of crop and animal selection, breeding and reproduction, and the conservation of existing genetic diversity.

**Box 7.5**
**Combating East Coast Fever**

Scientists at the Nairobi-based International Livestock Research Institute (ILRI) (1.) have developed an ELISA test with over 95% specificity against *Theileria parva*, the causative agent of East Coast Fever, a major tick-borne disease of African cattle. The test is currently undergoing field validation before being recommended for wider use. It is based on the isolation of a highly sensitive antigen known as the polymorphic immunodominant molecule (PIM). This was

transferred through genetic engineering to the bacterium *Escherichia coli*, then multiplied and used directly to coat the sensitive plates used in the ELISA test. Similar kits are being developed for other parasites, including *Theileria mutans, Anaplasma marginale* and *Babesia bigemina.* Polymerase chain reaction is being used to differentiate between *Theileria* species.

Working in collaboration with a private-sector laboratory, ILRI's scientists have also developed a trial vaccine to control East Coast Fever. This was based on the

identification of a gene coding for the P67 protein in *T. parva*, previously recognised as heavily targeted by antibodies and therefore a good candidate for a vaccine. Together, the ELISA test and the new vaccine provide a formidable set of new tools with which to combat one of sub-Saharan Africa's most devastating livestock diseases.

Source: ILRAD (1995).

1. Formerly the International Laboratory for Research on Animal Diseases (ILRAD), now merged with the Addis Ababa-based International Livestock Centre for Africa (ILCA) to form the new institute.

## Selection and breeding

A central theme in agriculture is the exploitation of natural genetic variability to improve the crop varieties and animal species available to producers. For centuries, this task has been the sole responsibility of farmers, but in the past 100 years the modern science of plant and animal selection and breeding has evolved, driven by the need to achieve higher and more stable yields to feed rapidly rising human populations.

Since the 1970s, several advanced techniques have been developed to aid the identification and transfer of desirable traits. In crops, these include tissue culture (anther and microspore culture and *in vitro* selection), and in both crops and animals, genetic fingerprinting and the use of genetic markers. In addition, germplasm can now be improved through several techniques that allow access to genepools previously too distant to be tapped through conventional breeding or in natural genetic exchange. These are the techniques of genetic modification, which are combined with those of *in vitro* cell and tissue culture. Finally, genetic diversity can be increased through various undirected approaches such as mutation induction and somatic hybridisation or cybridisation. These are suitable when directed approaches to obtain the desired end product are not feasible.

### Tissue culture

Tissue culture is a collection of techniques whose main application so far has been the mass propagation of planting materials. As such, it will be discussed in more detail in the section on Reproduction, below. We deal here with specific applications that aid in germplasm improvement.

### Anther and microspore culture

New methods have been developed for making better use of genetic variability in crops by bringing to light recessive traits. This can be achieved by trait analysis of pollen, followed by genome duplication and regeneration into plants. Pollen is abundantly available, as every flower has several anthers, each containing thousands of pollen grains. Whereas all other regeneration methods produce plants containing the same genetic information as the donor plant, pollen-derived plants provide a broader array of genotypes. The technique also has the advantage of achieving homozygosity in a single step, instead of after several years of backcrossing. Pollen and anthers carry haploid cells, containing only one copy of each chromosome, so chromosome duplication gives rise to fertile, diploid plants with two identical sets of chromosomes. This implies a considerable reduction in the time required for the selection of new varieties with desired characteristics, either from these homozygous diploids but more often from crosses exploiting these diploids, particularly in the case of highly heterozygous cross-pollinating species. If a crop is liable to inbreeding, the efficiency of the technique is drastically reduced.

So far, anther culture has been performed successfully for a large number of both monocotyledonous and dicotyledonous species. For pollen culture, immature pollen (microspores) is needed. This technique permits the use of single cells with unmasked recessive traits in applications such as *in vitro* selection, mutation induction and genetic modification (see below).

Many developing countries are applying anther culture in their conventional crop breeding programmes. This technique demands only modest facilities, the main requirement being the ability to work under sterile conditions, which involves using laminar flow benches. Microspore culture, however, is more complicated and requires more elaborate facilities, including a growth room and greenhouse.

### In vitro selection

In this technique, specific stress situations are simulated *in vitro,* leading to the direct selection of plant cells. The large number of individuals that can be screened in a limited space make this an attractive approach. Prerequisites for *in vitro* selection are that a suitable regeneration method for the plants must be available, and that one or more defined compounds must be known to play a role in stress induction at cellular level. There is a trade-off between regeneration and applied selection, the former benefitting from larger cell clusters, the latter from smaller ones. Protoplasts or single cells provide the best contact with the selecting agent, allowing more individuals to be screened, but regeneration is dramatically reduced.

*In vitro* selection can speed up the breeding of improved varieties, especially those requiring the incorporation of polygenic traits or unknown tolerance or resistance mechanisms. Its major limitation is that it is difficult to control and not at all well understood. With few exceptions, in cases where the selection conditions imposed at the cellular level ressemble those encountered by the plant under field conditions, this technique has not been very successful. It is, however, being used by many developing countries. The facilities required are similar to those for anther culture.

## Genetic fingerprints and markers

The genetic background of a plant or animal can be identified at the molecular level by making fingerprints of its DNA using more or less randomly selected DNA fragments, without the need to trace the genes responsible for specific traits. This technique, which normally employs relatively short sequences of DNA (10–20 nucleotides), allows samples of unknown origin to be characterised and compared, and genetic distances and relationships between different varieties to be assessed.

Genetic markers, in contrast, are often longer sequences of DNA which correlate with the presence or absence of a specific trait. They are primarily used to increase the efficiency of progeny screening in breeding programmes. As well as aiding in the introgression of single genes, they enable complex polygenic traits such as drought tolerance to be analysed. In addition, correlations between DNA markers and segregation patterns allow traits to be associated with specific chromosomes and chromosome regions, such that a map of the plant genome is gradually constructed. This in turn allows the selection of chromosome segments and the eventual isolation of genes (Box 7.6).

---

**Box 7.6**
***A lie detector for cassava breeders***

Two problems dog the improvement of cassava through conventional crop breeding. First, the plant's long time to maturity, 12 months or more, means that at least 10 years are needed to develop a new variety. Second, most of the genes responsible for important traits are recessive. As a result, many cassava plants look more similar than they really are, making traditional selection a game of blind man's buff.

To combat these problems, scientists at CIAT are developing a molecular map of cassava. The team began by crossing a high-yielding Latin American variety with a plant from Nigeria resistant to African cassava mosaic virus, a disease that causes extensive losses in sub-Saharan Africa. They then used restriction enzymes to cut the progeny of the cross into DNA fragments, forming a genomic "library". (This is really a misnomer, since the outcome at this stage is more like the jumbled pieces of a huge jig-saw puzzle). Next, segregating populations and markers were used to order the pieces into linkage groups – segments of DNA that are inherited together. During this phase the broad outlines of the map are obtained. The scientists are now in the final stage, in which different markers are tested for their correlation with specific traits to refine the map's accuracy and usefulness.

Making the map is a painstaking task of mammoth proportions, but the end result should bring significant gains in the speed and accuracy of germplasm characterisation and plant breeding. Breeders will be able to screen hundreds of genotypes a day, instead of only 10 at present. They will also be able to use the laboratory rather than having to grow plants out in the field. Better still, the map should make problems with recessive genes a thing of the past. Instead of having to depend on the plant's phenotype for selection, scientists will be able to delve beneath appearances to reach the genome. As one scientist put it: "the map will be our lie detector".

Source: CIAT (1995).

The DNA fragments used for fingerprinting or as markers are either applied in a radio-active form or with an enzyme attached that catalyses the generation of a fluorescent label. In both cases they are known as probes.

According to need and to the facilities available, three major types of marker can be used to characterise germplasm: restriction fragment length polymorphism (RFLP), amplified fragment length polymorphism (AFLP), and random amplified polymorphic DNA (RAPD). The RFLP technique is the oldest and most reliable, but it requires the use of restriction enzymes and elaborate facilities, including hybridisation equipment, X-ray film and cold storage. The newer techniques, RAPD and AFLP, show promise in applied breeding programmes in the long term, as they are quicker (Caetano-Anollés et al., 1991; Waugh and Powell, 1992). However, being less robust, they also require fine-tuned equipment and procedures. Both AFLP and RAPD rely on the polymerase chain reaction (PCR) technique, which implies the use of special equipment. Like the RFLP method, they also rely on gel electrophoresis to visualise the results. The time needed to elucidate a correlation between genetic marker and desired trait depends on the trait and on the extent and detail of the genetic map available, if any.

The use of genetic markers may be regarded as a useful step in applying advanced techniques to the characterisation and improvement of local crop varieties. However, this technique demands substantial investments in skills and facilities, and its effective use depends on clear objectives and strong commitment on the part of staff. Few developing countries have yet become extensively involved in the use of these technologies in their national programmes, but many enjoy access to them through joint projects with the international agricultural research centres of the Consultative Group on International Agricultural Research (CGIAR). For crops and animals of economic interest to the developed world, extensive research has been conducted and genomic maps have been established. On crops and animals restricted to developing countries and of minor economic importance to the developed world, rather less has been done, but research at the international centres has made good progress in recent years.

## Genetic modification

Genetic engineering or modification (3) is the direct introduction of DNA into a plant or animal cell, subsequently giving rise to an intact individual with a new trait. Prerequisites for the use of these techniques are that the gene which codes for the desired trait has been identified and that appropriate plant regeneration or animal embryo transfer protocols have been developed.

There are several techniques for introducing DNA into a plant cell, and each can be used with one or more regeneration methods. The earlier techniques could be applied to dicotyledons only. The most common approach is the use of a transfer vector (4) derived from the Ti-plasmid (tumor-inducing plasmid), which is found in the bacterium *Agrobacterium tumefaciens* (the causative organism of the crown-gall tumours that often affect wounded dicotyledonous plants). The Ti-plasmid can insert part of itself into the genome of most dicotyledonous plant cells. Unfortunately, only very few monocotyledonous species can be transformed using the Ti-plasmid.

The most promising approach for monocotyledons, in which relatively simple regeneration systems (including tissues or organs) can be used, is particle-mediated DNA delivery. In this technique, the DNA is coated on small particles (of for example gold or tungsten) of about 1 $\mu$m in diameter. These are then shot into the plant tissue or single cells, using a particle gun. The particle gun was first developed by Klein et al. in 1987. Since then, several improved or modified versions have been developed by other scientists, partly to circumvent the patents taken out. The apparatus can either be leased, which is rather expensive, or home-made (Finer et al., 1992). Many researchers have successfully applied this technique in cereals. The first cereal to be successfully genetically modified was maize in 1990 (Fromm et al.), followed by wheat, rice and barley.

A still more sophisticated approach, usually requiring the use of protoplasts, is that of electroporation. In this technique, embryonic cells first undergo mild enzyme treatment before being subjected to electrical pulses in the presence of the DNA for a period of milliseconds. This approach seems to be reproducible, but so far very few people have used it, on only a few crops (amongst which various oilseeds and maize). Knowledge of its usefulness is therefore limited compared to particle-mediated DNA transfer.

In animals, modification of the host cell is mainly achieved by micro-injection of DNA in the pronucleus (containing half the number of chromosomes) of the sperm cell just before it fuses with the pronucleus of the egg cell. Electroporation and virus infection have also been applied. Transformed cells, resulting from a fusion of male and female gametes and subsequent modification, must be manipulated using embryo transfer techniques to obtain genetically modified progeny.

Research on genetic modification in plants began in the early 1970s and led to the development of transfer vectors a decade later. The first foreign gene transfer occurred in 1983, when a bacterial gene coding for antibiotic resistance was transferred to petunia. Since then, scientists have successfully transferred single genes controlling various agronomically important traits such as resistance to viruses, insects and herbicides, and tolerance to toxic minerals, as well as earliness, delayed ripening and altered carbohydrate contents. More than 50 species, including cabbage, carrot, cotton, lettuce, rape, peas, potato, sugar beet, rice, maize, soybean, tobacco and tomato, have been genetically modified in this way. In the late 1980s, a technique for transferring the so-called antisense genes coding for antisense-RNA was developed. By deregulating the expression of certain genes, it has proved possible to change the colour of flowers (for example petunia) and to reduce post-harvest biodegradation (for example tomato ripening). Despite these technological advances, commercial applications of transgenic crops have so far been few, largely because of low public acceptance and slow market approval procedures in most developed countries. Recently, USA authorities have approved the marketing of a transgenic tomato with improved keeping quality. In China, tobacco plants with genetically engineered disease resistance have been released to farmers.

The first deliberate incorporation of a foreign gene into an animal genome came in 1980, when scientists transferred a gene coding for rat growth hormone into a mouse, which grew to a much larger size than normal. Similar genes were then transferred to various farm animals, including swine, sheep and cattle. Since 1982, hundreds of different genes, human and animal, have been inserted into a variety of mammals, birds and fish, not only to enhance the production of conventional animal products, such as meat, milk and wool, but also to produce novel products such as biomedical proteins (see Box 7.7). However, there have so far been very few commercial applications of genetic modification in animals, since the technologies of both transfer and regeneration are much more complicated than for most plants. One commercial but highly controversial application is the Harvard or cancer mouse, which has been engineered to have a disfunctional immune system, making it suitable for use in medical research but anathema to animal welfare activists.

Genetic modification has limited applicability and is still hampered by several serious technical problems. It is necessary only if the desired trait is not present in the available genepool of the species or in intercrossing relatives, and possible only if the gene responsible for it has been identified and isolated. The latter condition has still been met in only a limited number of cases. In addition, appropriate regeneration or embryo transfer protocols for use with genetic modification are essential. Even if transfer and regeneration are carried out successfully, it remains to be seen whether the introduced gene will be stably expressed in later generations. In some instances, it has been lost in subsequent breeding, although in most cases modification has proved stable over many generations. So far, only single-gene characteristics have been successfully transferred. Many important traits, such as yield potential, product quality and drought tolerance, depend on the interplay of multiple genes (polygenic traits), which have not yet been isolated and thus cannot be transferred. The success rate of transformation with *Agrobacterium* is relatively high (about 30%), but that of other techniques is much lower, often below one per cent. The mechanism by which foreign DNA is taken up and incorporated into the host genome remains largely unknown, and the site of incorporation cannot be targeted.

---

**Box 7.7**
**Herman the transgenic bull**

Cows are highly suitable for the production of biomedical proteins. They produce 10 000 litres of milk a year, containing 330 kg of proteins. Proteins can be easily purified from the milk.

Researchers of the Dutch-based biotechnology firm, Gene Pharming Europe, have successfully transformed cattle with a view to harnessing this aptitude. In their laboratories, oocytes are taken from ovaria and are then matured and fertilised *in vitro*. They are then micro-injected with foreign DNA, before the two pronuclei are fused. After seven days, the embryo is transferred to an acceptor cow.

Using this method, a transgenic bull named Herman was born in 1990. Herman carries a gene coding for human lactoferrin, a protein excreted in mother's milk that retards the growth of bacteria. Whether the experiment is a scientific and commercial success remains to be seen, since fertility problems have arisen and the effectiveness of lactoferrin as a medicine has yet to be demonstrated. However, the experiment aroused considerable public discussion in the Netherlands on the ethics of applying genetic modification to domestic animals.

Glastra van Loon and Kuiper (1995).

The advantage of genetic modification is that it allows access to genes in species wholly unrelated to the one being altered and hence inaccessible using conventional breeding techniques. It is, however, much more expensive than conventional breeding, and does not necessarily reduce the amount of time needed to develop a new crop variety or animal breed (although it does obviate the need for time-consuming back-crossing). Moreover, genetic modification requires sophisticated facilities and skills, including specialised maintenance of expensive equipment. Both researchers and support staff need to be highly trained. Developing countries establishing their own in-house capacity for such research must purchase most of the necessary equipment, spare parts and chemicals abroad. Delivery times can be long and shortages of hard currency may arise at any time.

Most genetic modification research is located in the private sector of developed countries. Traits of interest to Western agriculture, such as herbicide tolerance, insect and virus resistance and quality characteristics, have been extensively researched. Thailand, India, Brazil, Mexico and Zimbabwe are among the few developing countries that have so far developed a national capacity for this kind of research. However, many others have been able to join networks or form research partnerships with the international agricultural research centres, most of which have already done some work on genetic modification in at least some of their mandate crops. Examples are the International Crops

Growth room used for micropropagation of various crops at ICRISAT, Hyderabad, India
(*Rommert van den Bos*).

Research Institute for the Semi-Arid Tropics (ICRISAT), working on sorghum, the Centro Internacional de Mejoramiento de Maïz y Trigo (CIMMYT), on maize, the Centro Internacional de Agricultura Tropical (CIAT), on bean, the Centro Internacional de la Papa (CIP), on potato, the International Rice Research Institute (IRRI), on rice, and the International Institute of Tropical Agriculture (IITA), on cowpea (Box 7.8). Research on genetic modification is expensive and complicated, and progress in crops of interest to small-scale farmers in developing countries has so far been limited, but this does not mean that such research has nothing to offer to small-scale agriculture. Where small-scale farmers regularly purchase seeds of a specific crop on the market instead of retaining their own seed, the addition of traits such as resistance to specific diseases to existing, well-adapted varieties may well be of interest. A prerequisite for success will be that the often higher price of new seed is justified in the eyes of the farmers by reliable yield increases.

## Mutation induction

Genetic variation can be effected via the mutation of plant cells or seeds induced by chemical compounds or by irradiation. Mutation induction is easily applied, but requires strict safety measures. Its major disadvantage is that the use of mutagens may result in a product that has undergone undesirable hidden mutations besides those intended. It is essential that the desired trait can be easily detected, as in most cases thousands of plants have to be screened because of low success rates. An advantage of this technique is the ability to screen at the cellular level prior to regeneration. The time required for mutation induction depends on the character desired and the nature of the screening involved.

---

**Box 7.8**
**Strengthening insect resistance in cowpea**

Cowpea (*Vigna unguiculata*) is a drought-tolerant legume vital to the food security and livelihoods of poor people in sub-Saharan Africa. Conventional breeding research at IITA has already led to the release of several higher yielding, short-duration varieties. But the new varieties' Achilles heel is their susceptibility to insect pests, particularly pod-suckers, thrips and the *Maruca* pod-borer, which can decimate farmers' yields.

Working in collaboration with national partners and laboratories in developed countries, IITA's scientists are tackling the problem by accessing genes in wild *Vigna* and other species only distantly related to the modern cowpea found in farmers' fields. In 1991 the first transformed cowpea embryos were obtained, using a particle gun for gene transfer. Transformation using *Agrobacterium* as a vector was also achieved in the same year.

To support the effort, IITA has broadened its germplasm collection to include more than 500 accessions of wild *Vigna* species. Many of these are from southern Africa, a previously neglected centre of genetic diversity for cowpea. Genetic markers are being used to characterise accessions and identify genes conferring resistance. Work is under way to construct a DNA map of cowpea to aid future breeding efforts.

Source: IITA (1992).

Another approach to increasing genetic variability is through mutations arising spontaneously from *in vitro* culture (somaclonal variation). The frequency of mutation is lower than when mutagens are used, and not all plants are equally susceptible (Karp et al., 1987). Consequently, even more plants have to be produced and screened. Moreover, genotypic changes are not always stable through generations when this technique is used. One of the few successful applications of this approach was the production of plants tolerant to downy mildew (*Sclerospora graminicola* Sacc.) in pearl millet. In most laboratories in industrial countries, the technique is seldom used nowadays due to its apparent ineffectiveness.

Mutation induction is a relatively simple technique that can easily be applied in developing countries by any laboratory possessing tissue culture facilities. When regeneration methods and facilities are also available, mutation induction can be considered an option in crop breeding if no alternative solution exists to an urgent problem, but it should be realised that its outcome is highly unpredictable.

### Somatic hybridisation and cybridisation

Most plants contain genetic information both in the cell nucleus and in the cytoplasm. Protoplasts can be fused (somatic hybridisation), yielding a fusion product which initially contains the entire nuclear and cytoplasmic genetic information of the two cells. Subsequently, nuclei fuse and genomes reassort and recombine, producing a wide variety of gene combinations. The regenerated plants have characteristics of both parents. Selective destruction of either nuclear or cytoplasmic DNA will produce a fusion product which can combine the nucleus-borne information of the one parent with the cytoplasm-borne information of the other (cybridisation). Using this method, organelle-encoded traits such as cytoplasmic male sterility (important for the production of hybrid varieties) or disease resistance can be transferred unilaterally. Furthermore, the hybridisation method can be used to overcome the crossing barriers that often impede conventional breeding. A curious example is the viable crosses achieved between potato and tomato. Some disease resistance traits have been bred into important crop species, especially potato and tomato, using somatic hybridisation.

However, the availability of an appropriate selection method for the desired trait as well as a reliable protoplast culture system is critical. The latter is available for a few crops only at present. In most cases, somatic hybridisation and cybridisation do not prove very useful, and the techniques of genetic modification are usually preferred nowadays.

### Reproduction

Once an improved crop variety or animal breed has been developed, it becomes essential to provide farmers with sufficient cheap and high-quality stock of it. In this section we will examine the various science-based biotechnologies that can assist in plant and animal reproduction: mass propagation of planting material, embryo technology, and technology for detecting hormone levels in animals. (The latter is indicative of their reproductive potential.)

## Mass propagation of planting material

Plants can be regenerated from organs (leaf, stem, embryo, anther), single cells (including microspores), cell clusters (callus), or from cells without a cell wall (protoplasts). The lower the level of organisation in the source material, the more difficult it is to induce differentiation in the full plant. In other words, it is easier to regenerate plants from a leaf than from protoplasts. Furthermore, the more difficult it is to induce the necessary differentiation, the longer the time required for *in vitro* culture. This increases the risk of random, uncontrolled mutations and generally reduces regeneration potential.

Planting material can be mass propagated through tissue culture. This technique is well suited for use with crops that cannot easily be propagated by traditional means, such as many slow-growing woody species, species with low fertility, or highly heterozygous crop varieties. Tissue culture can also be used to multiply a new variety rapidly. A group of crops for which the technique appears particularly attractive are vegetatively propagated crops that are highly susceptible to viral diseases. Such diseases are usually absent or occur at greatly reduced concentration in the meristem tip of plants. Meristem culture, in which this tip is isolated and used for cultivation *in vitro*, is a form of tissue culture that can be used to eliminate viral diseases. Combined with initial high-temperature growing conditions, which add further protection against viruses, tissue culture has now been applied to over 50 plant varieties, including many ornamentals, strawberry, cocoa, grape, lemon, the important root and tuber crops such as potato, cassava, sweet potato and yam, and several commercially important tree species, such as oil palm and banana.

The use of tissue culture is much simpler for dicotyledons than for the notoriously recalcitrant monocotyledons, which are more difficult to regenerate in this way. However, for many useful crop species tissue culture represents a straightforward technique which most developing countries have already mastered. Its application requires no more than a sterile working place (a horizontal laminar air-flow hood), a growth room and a simple greenhouse, and some (not necessarily highly) trained manpower. Tissue culture is labour-intensive and consequently rather costly in industrial countries, where plants produced in this way must usually be sold for more than US$ 0.35 each if costs are to be recovered, and unit prices of over US$ 1.00 are normal for most plants. Despite lower labour costs in developing countries, the technique is still likely to be a relatively expensive option in the production of large volumes of plants, especially when the planting material is seed. However, several countries report the successful commercial use of tissue culture propagation for the production of early-generation, disease-free plants, and some high-value plants in horticulture. In India, for example, tissue culture banana plantlets can be obtained for about US$ 0.45 each. Box 7.9 describes the mass propagation of potato in Vietnam.

A common practical problem in tissue culture is the inability to maintain sterile conditions, leading to contaminated planting material. Another, more far-reaching consequence is the genetic uniformity that may result when only one genotype is used for mass propagation. Two cautionary tales warrant mention here. First, a project to mass propagate yam on the Caribbean island of Barbados failed, due partly to difficulties in maintaining high hygienic standards over the years and partly to the selection and use of lines that proved highly susceptible to anthracnose, a major disease in the region (Broerse et al., 1991). Second, an oil palm propagation project of Unilever in Malaysia encountered a severe setback as it entered mass production, when the flowering of the tree, necessary for oil production, became abnormal because culture conditions had been simplified, inducing somaclonal variation (Walgate, 1990).

## Detection of hormone levels

Monoclonal antibody technology is used not only in disease diagnosis but also in the detection of hormone levels in breeding female animals. An example is the milk progesterone test, an ELISA which can be used to determine the progesterone content in milk samples of cows or female buffaloes. The test is applied to

---

**Box 7.9**
**Mass propagation of potato in Vietnam**

A new potato variety resistant against the fungus *Phytophtora infestans* was introduced in Dalat, Vietnam in 1979. After on-farm trials, a plan for further dissemination was developed. It was decided to establish local small-scale enterprises to handle both the production and distribution of planting materials. Financed by the Vietnamese Government, the project started in 1981, when 10 farmers established commercial production. Multiple shoots were bought from a central laboratory and transplanted into sterile sand for rooting. The plantlets were sold to farmers at a profit. Although buoyant during the first year, demand fell considerably in subsequent years, since farmers wished to retain their own seed potatoes for as long as possible before buying new planting material. As a result, only three out of 10 production units were still operating after five years.

The partial success of the project encouraged the government to initiate similar projects in other potato-growing areas in Vietnam. However, these attempts were much less successful. Climate, soil fertility and crop management were less suitable for the use of tissue culture plantlets. There was little demand for the plantlets and the production units ran at a loss.

This example demonstrates the importance of analysing the market before introducing a biotechnology-based innovation, however simple that innovation may be.

Source: Broerse and Stokhof (1993).

find out whether an animal is in heat, and also to check for pregnancy. It is mainly used as an aid in providing timely artificial insemination services and in reducing calving intervals, which can be too long if oestrus and pregnancy are misjudged by farmers. But it can also be used as an investigative tool in the analysis of the causes of infertility. A "cow-side" version of the test has been developed for application by farmers (Booman et al., 1990), the results of which are available within five minutes.

Many developing countries have already mastered this simple and robust technique, which is, however, mainly applied in combination with artificial insemination. With the exception of some producers in South-East Asia and East Africa, most small-scale farmers in developing countries do not have access to artificial insemination. Consequently, this diagnostic test is mainly of interest to intensive large-scale milk production units around major cities.

## Embryo technology

Over the past 25 years, techniques for the recovery, storage, transport and implantation of animal embryos (embryo transfer, or ET) have been developed to a stage at which they are practically applicable in cattle, sheep and goat breeding (Persley, 1990). ET is still not routinely possible in swine and buffalo. Nowadays, embryo transfer is often combined with super-ovulation, induced by the follicle-stimulating hormone (FSH), and with *in vitro* fertilisation. The principal benefit of ET is the ability to produce more offspring from an elite female animal than would otherwise be possible. For example, a cow will give birth to about four calves in an average lifetime. With ET, this can be increased to at least 25 calves. The technique could thus vastly speed up the spread of rare genetic stock (for example, a new breed). ET can also reduce the costs of importing cattle, since importing animals as embryos is much cheaper. Moreover, raising these animals in their new home country makes them better adapted to local environmental conditions when adult (Xu et al., 1987; Persley, 1990).

There are other potential advantages from ET. Following *in vitro* fertilisation, a single embryo can be cloned through mechanical splitting or transplantation of nuclei, resulting in two or more genetically identical offspring. Another option is embryo sexing, which could further increase selection intensities. For example, it could permit greater specialisation of the beef and milk production functions of a dual-purpose population. Furthermore, ET is an essential step in the genetic modification of improved animals. A more esoteric application is to fuse meiotic cells to create chimaeras (5) such as the "geep" or "shoat", a cross between a sheep and a goat.

The costs of ET are high. In developing countries, the hormones involved, and often the embryos themselves, have to be imported. The application of ET in the field demands access to proper methods for cryopreservation, so that embryos collected at an experiment station can be transported over long distances. Costs per embryo transferred have been estimated at around US$ 250 (Jansen, 1991). In many countries in Asia and Latin America, there have been experiments with ET in cows and buffaloes. The work on cows has in the main been quite successful. So far, field applications in developing countries have been restricted to large-scale livestock enterprises. ET is unlikely to be appropriate at the small-scale farm level.

## Germplasm conservation

Most of the world's genetic diversity is found in the tropics. Traditional agriculture in the developing countries preserved the diversity of crop and other genetic resources, but the widespread adoption of improved varieties, combined with such pressures as deforestation and soil erosion, has threatened many species with extinction, especially landraces and their wild relatives. These plants may be better adapted to locally dominant stresses than so-called improved varieties. They may also harbour important genes for future crop breeding efforts. For these reasons both *ex situ* conservation in gene banks and *in situ* conservation in farmer's fields have been pursued. These approaches are discussed in Chapter 6.

In *ex situ* conservation, stored material is usually grown out at regular intervals to ensure its continuing viability and adaptation to the changing external environment. For those species that grow slowly, have low fertility or do not remain viable under current storage conditions, tissue culture can be an effective alternative way of propagating selected genotypes.

A major problem of both *in situ* and *ex situ* conservation is the sheer size that collections can acquire. There is a need for rational sampling methods to ensure that the widest possible range of genetic diversity is included in collections that remain of manageable size. Biotechnology offers new methods of assessing genetic diversity and the relationships between different populations or samples. Especially useful is the analysis of isozymes, which can be done easily, rapidly and at relatively low cost. The use of this technology may serve the same purposes as genetic fingerprinting, the former technology being older and simpler to apply while the latter is more sensitive.

On the livestock side, cryopreservation of animal embryos is still in its infancy. Much research is still needed before a system of gene banks for animals equivalent to that for crops is possible.

Currently, most genebanks are public institutions. Their establishment and maintenance requires considerable investment in facilities and skilled staff. Since the benefits of such investment are felt only in the long term, genebanks tend to get low priority from most governments in developing countries. This underscores the important part played in this field by the international agricultural research institutes of the CGIAR, which collect and store landraces and other genetic material of most of the world's important food crops. This effort does not obviate the need to maintain landraces in their original environments, subject to continuing adaptation and selection by farmers, wherever possible (see Chapter 6).

## Conclusions

Without looking in detail at the local circumstances of small-scale farmers in developing countries, it is impossible to predict in advance which of the wide range of science-based biotechnologies now available will be of benefit to them (Broerse and Bunders, 1991). Like any other technological innovation in agriculture, a biotechnological innovation may be highly appropriate in one region but completely useless in another. Some general observations can nonetheless be made. First, biotechnologies are usually more expensive than conventional technologies, with the result that fewer of them will be affordable for resource-poor farmers. In these circumstances it is even more important to get the answers right concerning relevance to farmers' needs before investing money in research and extension. Second, many biotechnologies benefit researchers and the research process, its speed and efficiency, rather than being directly usable by farmers. In such cases the benefit of the technology will depend on its applications, not on the technology itself. New applications are hard to predict in such a fast-changing field.

Science-based biotechnology is a highly dynamic field in which prospects can change rapidly. Generally, however, the over-optimism and exaggerated claims that characterised its early advances have now given way to a more sober assessment of its potential, accompanied by a better understanding of what techniques will and will not prove useful in developing country settings.

Some biotechnologies are already widely used in developing countries. All countries use traditional fermentation processes, especially solid-state, alcoholic and lactic acid fermentations, and most have successfully introduced at least some modern large-scale processes alongside the traditional sector, even if only for brewing beer. Most countries have also mastered plant tissue culture, although mainly in the laboratory rather than on a commercial scale. In addition, biofertilisers and a range of biological control agents, notably *Bacillus thuringiensis,* are now being produced in some countries. Improving research and development capabilities in all these fields should be neither very complex nor very expensive.

Such fields as the production of high-quality enzymes, diagnostics and vaccines, and the use of genetic markers, are technically more complicated. Most developing countries cannot yet efficiently mobilise any of these technologies on a significant scale, although laboratory-level production is often feasible. The application, as opposed to the production, of enzymes, diagnostics and DNA markers is relatively easy. For the time being, the best option for most developing countries is to obtain these technologies through close collaboration with laboratories, institutions or companies, mostly located in the developed countries, that have the necessary expertise and equipment.

Genetic modification is the most complex and expensive of the science-based biotechnologies. Furthermore, it carries potential risks, not merely for the health and safety of immediate users but also for society as a whole, since the environmental impact of organisms released in new forms cannot be fully known in advance. With the exception of a few university laboratories and research institutes in India, Thailand, Brazil, Mexico, Colombia, Zimbabwe and Kenya, the national research systems of most developing countries do not yet have the capacity to apply genetic modification, and for most it may not be realistic to seek to develop such a capacity at national level in the near future. Most countries also lack trained scientists in the basic disciplines required for genetic modification.

For the many countries that have yet to embark on genetic modification, the best option is probably to tap the expertise and equipment available at the international agricultural research centres, and at institutes in the more technologically advanced countries in their region, and, possibly, in the developed countries. The institutes of the developed world may have the most advanced research capabilities, but they rarely focus on the crops and problems of developing countries. The international centres, on the other hand, are in many cases based in the developing countries, and have long had mandates and research activities geared towards meeting the needs of small-scale farmers. The challenge for national research systems is to exploit the capacities of these centres, adapting the materials and technologies they produce to suit their own needs.

The release of genetically engineered organisms is an important issue which needs further research and regulation. Many industrial countries have imposed strict controls in this area. In most of these countries risk assessment is conducted largely on a case-by-case basis, but certain applications in specific crops are now evaluated along generic lines. With the notable exceptions of India and the Philippines, no developing countries currently have legislation on this subject, although several (Zimbabwe, Kenya, and a number of Latin American countries) are taking steps in this direction. Even once regulations have been introduced, it will be difficult to police them due to lack of expertise and infrastructure. Extensive staff training, more experience and regional co-operation are urgently needed in this area.

Biotechnologies are as likely as other science-based technologies to suffer the fate of non-adoption by those for whom they

*Biotechnology, and especially...genetic engineering, is both capital- and equipment-intensive...The equipment is increasingly going beyond the reach of most Third World countries as well as public-sector institutions.*    Vandana Shiva, at the Conference of Parties to the Convention on Biological Diversity, Jakarta, 1995.

were intended. Countless adoption studies have shown that innovations once hailed as major solutions have proved not to be attractive to farmers after all, for a variety of reasons including inappropriateness to the farming system or agro-ecology, and high cost and/or low profitability. Many technologies are designed to increase yields under optimal conditions rather than to stabilise them under stressful or unpredictable conditions, a characteristic more useful to risk-averse resource-poor farmers in marginal environments. Other technologies have simply not addressed the most pressing problems of farmers, or have addressed only those of progressive male farmers. Those that have been adopted have sometimes had unforeseen side-effects, such as an increase in the (unpaid) workload of women or a harmful impact on the environment. The practitioners of science-based, public-sector biotechnology research and development must learn these lessons from the past and seek, through new research partnerships, a better fit between the immense power of the technologies at their disposal and the very great, but often highly location-specific, needs of the small-scale producers and consumers whose needs they serve.

In sum, provided a suitable research paradigm can be developed, one that succeeds in incorporating a user orientation into the design of projects that must, by definition, involve research in the laboratory alongside that on farmers' fields, science-based biotechnologies may gradually come to benefit the many millions of small-scale farmers who will increasingly look to science for solutions to the pressing food production and environmental problems of tomorrow.

## Notes

(1)  When a foreign entity (antigen) enters the circulatory system of a higher vertebrate, it stimulates a specific group of white blood cells, the lymphocytes, to produce antibodies. One lymphocyte produces one type of antibody. Antibodies combine with the antigen to cause its destruction within, or removal from, the body. An antigen may possess several determinants, inducing the formation of several different antibodies.

(2)  MIRCENs are existing academic and/or research institutes in developed and developing countries. These centres participate in a global collaborative network to promote beneficial applications of micro-organisms through research and development.

(3)  The two terms are synonomous, but the fact that the latter is now more widely used is symptomatic of the public mistrust for this technology. In the early days, scientists proudly referred to genetic engineering, stressing its difference from conventional breeding techniques. Latterly they have adopted the vaguer genetic modification in an attempt to blur that difference.

(4)  A vector can be a plasmid (circle of extrachromosomal DNA), virus, or transposon (segment of DNA which can move from place to place within a genome). In all cases, the vector is used to help the foreign gene move into and replicate within the new host. The final construct

contains certain control sequences which ensure that the gene, once integrated into the host genome, is correctly transcribed into RNA and translated into a protein.

(5) The genetic material of the two parental embryos remains separate but is dually expressed in the same animal. The resultant offspring consists of a mosaic of cells, some of which carry genes from one parent and some from the other.

## References

Ahmed, S.M. and Ahamad, A. 1992, 'Efficacy of some indigenous plants and pulse protectants against *Callosobruchus chinensis* (L.) infestation,' *International Pest Control*, March/April: 54–55.

Anonymous, 1993, 'US authorities give approval for Finnish biological fungicide,' *International Pest Control* July–August: 102

Antebi, E. and Fishlock, D. 1985. *Biotechnology: Strategies for Life,* MIT Press, Cambridge, Massachusetts and London, UK

ATI, 1988 'Rhizobium inoculant in Thailand,' ATI Bulletin 15, Washington DC, USA

Baker, R. 1989, 'Improved *Trichoderma* spp. for promoting crop productivity,' *Tibtech* 7: 34–38

Barnett, W.W., Edstrom, J.P., Coviello, R.L. and Zalom, F.P. 1993, 'Insect pathogen controls peach twig borer on fruits and almonds,' *California Agriculture* 47 (5): 4–6

Beudeker, R.F. 1990, 'Ontwikkeling van een microbieel fytase voor toepassing bij varkens en pluimvee (Development of microbial phytase for application in pig and poultry),' in A.W. Jongbloed and J. Coppoolse (eds), *Mestproblematiek: Aanpak via de Voeding van Varkens en Pluimvee: Proceedings of a Workshop on Animal Feed and Environmental Pollution*, Lelystad, 19 April 1990, pp.45–48

Booman, P., Tiemen, M., van Zaane, D., Bosma, A.A. and de Boer, G.F. 1990, 'Construction of a bovine-murine heteromyeloma cell line: Production of bovine monoclonal antibodies against rotavirus and pregnant mare serum gonadotrophin,' *Veterinary Immunology Immunopathology* 24, pp.211–226

Bos, K.D. 1990, 'Chemische achtergronden van fosforverbindingen en fytase in veevoeder (Chemical background of phosphorus compounds and phytase in animal feed),' in A.W. Jongbloed and J. Coppoolse (eds.), *Mestproblematiek: Aanpak via de Voeding van Varkens en Pluimvee: Proceedings of a Workshop on Animal Feed and Environmental Pollution,* Lelystad, 19 April 1990, pp.7–17

Broerse, J.E.W. and Bunders, J.F.G. 1991, 'The potential of biotechnology for small-scale agriculture,' in J.F.G. Bunders and J.E.W. Broerse (eds), *Appropriate Biotechnology in Small-scale Agriculture: How to Reorient Research and Development.* CAB International, Oxford, UK, pp.1–22

Broerse, J.E.W., Joffe, S. and Bunders, J.F.G. 1991, 'Towards criteria for assessment of project proposals,' in J.F.G. Bunders and J.E.W. Broerse (eds), *Appropriate Biotechnology in Small-scale Agriculture: How to Reorient Research and Development,* CAB International, Oxford, UK, pp.25–69

Broerse, J. and Stokhof, E. 1993, 'Milieu et industrialisatie (Environment and industrialisation),' in D. Nieuwenhuis, M. Priester and B. Wams (eds), *Samen of Niets: Naar een Politieke Keuze voor Wereldwijde Duurzame Ontwikkeling.* Evert Vermeer Stichting, Amsterdam, Netherlands, pp.55–63

Caetano-Anollés, G., Bassam, B.J. and Gresshoff, P.M. 1991 'DNA amplification fingerprinting using very short arbitrary oligonucleotide primers,' *Biotechnology* 9: pp.553–557

Campbell, R. 1989, 'The use of microbial inoculants in the biological control of plant diseases,' in R. Campbell and R.M. MacDonald (eds), *Microbial Inoculation of Crop Plants*. Special publications of the Society for General Microbiology, Vol.25, IRL Press, Oxford, UK, pp.67–77

Chandler, H., Kulkarni, S.G. and Berry, S.K. 1991 'Effectiveness of turmeric powder and mustard oil as protectants in stored milled rice against the rice weevil, *Sitophilus oryzae*,' *International Pest Control*, July/August: 94–97

CIAT, 1995, 'CIAT in perspective', Annual Report, Centro Internacional de Agricultura Tropical, Cali, Colombia

Clark, N. and Juma, C. 1991, *Biotechnology for Sustainable Development: Policy Options for Developing Countries*, ACTS Press, Nairobi, Kenya

Commandeur, P. and Komen, J. 1992, 'Biopesticides: Options for biological pest control increase,' *Biotechnology and Development Monitor* 13 (December): 6–7.

Dalzell, H.W., Biddlestone, A.J. and Gray, K.B. 1987, 'Soil management: Compost production and use in tropical and subtropical environments', FAO Soils Bulletin 56. Food and Agriculture Organisation, Rome

Das, H.K. 1991 'Biological nitrogen fixation in the context of Indian agriculture', *Current Science* 60: 551–555

De Jaeger, P. 1989, 'Zonder kunstmest (Without chemical fertiliser)', in *Natuur en Techniek* 10: 814–818

DiTomaso, J.M., Stowe, A.E. and Brown, P.H. 1993, 'Inhibition of lipid synthesis by diclofop-methyl is age-dependent in roots of oat and corn', *Pesticide Biochemistry and Physiology* 45: 210–219

Eaglesham, A.R.J. 1989, 'Global importance of rhizobium as an inoculant', in R.Campbell and R.M. Macdonald (eds), *Microbial Inoculation of Crop Plants,* Special publications of the Society for General Microbiology, Vol. 25, IRL Press, Oxford, UK, pp.29–48

Ely, S. 1993, 'The engineering of plants to express *Bacillus thuringiensis* d-endotoxins', in J.S. Corey, M.J. Baily and S. Higgs (eds), *Bacillus Thuringiensis: An Environmental Biopesticide*. John Wiley and Sons Ltd., Chichester, UK, pp.105–110

Feresu, S. 1992, 'Fermented milk products in Zimbabwe', in *Applications of Biotechnology to Traditional Fermented Foods,* Board on Science and Technology for International Development (BOSTID)/National Academy Press, Washington, DC, USA, pp.80–88

Finer, J.J., Vain, P., Jones, M.W. and McMullen, M.D. 1992, 'Development of the particle inflow gun for DNA delivery to plant cells', *Plant Cell Reports* 11: 323–328

Fromm, M.E., Morrish, F., Armstrong, C., Williams, R., Thomas, J. and Klein, T.M. 1990, 'Inheritance and expression of chimeric genes in the progeny of transgenic maize plants,' *Biotechnology* 8: 833–839

Giller, K.E., McDonagh, J.F. and Cadisch, G. 1994, 'Can biological nitrogen fixation sustain agriculture in the tropics?' in J. K. Syers and D.L. Rimmer (eds), *Soil Science and Sustainable Land Management in the Tropics,* CAB International, Oxford, UK, pp.173–191

Glastra van Loon, K. and Kuiper, K. 1995, *Herman: De Biografie van een Genetisch Gemanipuleerde Stier*, L.J. Veen, Amsterdam Antwerpen

Grace, J.K. and Yates, J.R. 1992, 'Behavioural effects of neem insecticide on *Coptotermes formosanus* (Isoptera: Rhinotermitidae)', *Tropical Pest Management* 38 (2): 176–180

Greenshields, R. and Rothman, H. 1986, 'Biotechnology and fermentation technology', In: S. Jacobssen, A. Jamison and H. Rothman (eds), *The Biotechnological Challenge*, Cambridge University Press, pp.77–95

Heiny, D.N. and Templeton, G.E. 1993, 'Economic comparisons of mycoherbicides to conventional herbicides', in J. Altman (ed.), *Pesticide Interactions in Crop Production*, CRC Press, pp.395–407

Hobbelink, H. 1991, *Biotechnology and the Future of World Agriculture*. Zed Books, London, UK

Hoben, H.J. and Somasegaran, P. 1992, 'A small glass fermentor for production of *Rhizobium* inoculum', NifTAL Project-MIRCEN: Illustrated Concepts in Agricultural Biotechnology No. 2. *World J. of Microbiology and Biotechnology* 8: 333–334

IITA, 1992, Annual Report 1991, International Institute of Tropical Agriculture, Ibadan, Nigeria

ILRAD, 1995, Annual Report 1994, International Laboratory for Research on Animal Diseases, Nairobi, Kenya

Jansen, J. 1991, 'Embryoproduktie en embryomanipulatie in een foktechnisch perspectief (Embryo production and manipulation from a breeding perspective)', in E. Egberts, T. van der Lende and A.J. van der Zijpp (eds), *Biotechnologie in de Veehouderij* (Biotechnology in Animal Production), Pudoc Wageningen, Netherlands, pp.47–52

Jovetic, S. 1994, 'Pyrethrins and production by *in vitro* systems', Directorate General for International Cooperation, Ministry of Foreign Affairs, The Hague, Netherlands

Karp, A., Steele, S.H., Parmar, S., Jones, M.G.K. and Shewry, P.R. 1987, 'Relative stability among barley plants regenerated from cultured immature embryos', *Genome* 29: 405–412

Klein, T.M., Wolf, E.D., Wu, R. and Sanford, J.C. 1987, 'High-velocity microprojectiles for delivering nucleic acids into living cells', *Nature* 327: 70–73.

Kloepper, J.W., Lifshitz, R. and Zablotowicz, R.M. 1989, 'Free-living bacterial inocula for enhancing crop productivity', *Tibtech* 7: 39–44

Koziel M.G. 1993, 'Field performance of elite transgenic maize plants expressing an insecticidal protein derived from *Bacillus thuringiensis*,' *Biotechnology* 11 (2): 194–200

Kuruvinakunnel, K.T.T. 1994, 'The modified anaerobic composting system', *ILEIA Newsletter* 10 (3): 16–17

Lynch, J.M. and Hobbie, J.E. 1988, 'Biological control', in J.M. Lynch and J.E. Hobbie (eds), *Micro-organisms in Action: Concepts and Applications in Microbial Ecology*, Blackwell Scientific Publications, pp.276–283

MacDonald, R.M. 1989, 'An overview of crop inoculation', in R. Campbell and R.M. MacDonald (eds), *Microbial Inoculation of Crop Plants*, Special publications of the Society for General Microbiology, Vol. 25, IRL Press, Oxford, UK, pp.1–9

MacKenzie, D. 1988, 'Science milked for all it's worth', *New Scientist*, 24 March, pp.28–29

McDonald, D. 1991, 'Biopesticides: Pesticides with a bright future?' *Integrated Pest Management* 2: 33–41

Mudgett, R.E. 1986, 'Solid state fermentation', in A.L. Demain and N.A. Solomon (eds), *Manual of Industrial Microbiology and Biotechnology*, American Society of Microbiology, Washington DC, USA, pp.66–83

Muscoplat, C.C. 1989, 'Commercialization and research perspectives for vaccines', in J. Cohen (ed.), *Strengthening Collaboration in Biotechnology: International Agricultural Research and the Private Sector*, Proceedings of a Conference held April 17–21, 1988 in Rosslyn, United States Agency for International Development (USAID), Washington, DC, USA, pp.141–149

National Research Council, 1992 *Neem: A Tree for Solving Global Problems,* National Academy Press, Washington, DC, USA

Nazareth, J. 1994, 'Micronutrient fortified compost,' *ILEIA Newsletter* 10 (3): 18–19

Netzer, W.J. 1987, 'Breakthrough in bio-insecticide research means transgenic plant technology must now be re-evaluated', Agricultural Genetics Report 6 (5), Mary Ann Liebert Inc., New York, USA

Nout, M.J.R. 1992, 'Upgrading traditional biotechnological processes', in *Applications of Biotechnology to Traditional Fermented Foods* Board of Science and Technology for International Development (BOSTID)/ National Academy Press, Washington, DC, USA, pp.11–19

Okafor, N. 1992, 'Commercialization of fermented foods in sub-Saharan Africa', in *Applications of Biotechnology to Traditional Fermented Foods,* Board of Science and Technology for International Development (BOSTID)/National Academy Press, Washington, DC, USA, pp.165–169

Persley, G.J. 1990, *Beyond Mendel's Garden: Biotechnology in the Service of World Agriculture,* CAB International, Oxford, UK

Postgate, J. 1990, 'Fixing the nitrogen fixers,' *New Scientist,* 3 February: 57–61

Primrose, S.B. 1987, *Modern Biotechnology,* Blackwell Scientific Publications, Oxford, UK

Rai, S.N., Singh, K., Gupta, B.N. and Walli, T.K. 1988, 'Microbial conversion of crop residues with reference to its energy utilisation by ruminants: An overview', in K. Singh and J.B. Schiere (eds), *Fibrous Crop Residues as Animal Feed,* Proceedings of an International Workshop held 27–28 October 1988, Bangalore, India

Reeves, J. 1991 'Integrated pest management reduces insecticide use by 80%', *Agriculture and Environment International* 45 (5/6): 57

Ruivenkamp, G. 1989 *The Introduction of Biotechnology in the Agro-industrial Production Chain,* Jan van Arkel Publishers, Utrecht, Netherlands

Salama, H.S. and Morris, O.N. 1993, 'The use of *Bacillus thuringiensis* in developing countries', in J.S. Corey, M.J. Bailay and S. Higgs (eds), *Bacillus Thuringiensis: An Environmental Biopesticide,* John Wiley and Sons Ltd., Chichester, UK, pp.237–253

Skerrit, J.H., Hill, A.S., Andrews, J.L., Edwards, S.L., Beasley, H.L. and McAdam, D.P. 1993, 'Application of immunological methods of cereal analysis', in J.R.N. Taylor, P.G. Randall and J.H. Viljoen (eds), *Cereal Science and Technology: Impact on a Changing Africa,* Papers from the ICC International Symposium. CSIR, Pretoria, South Africa, pp.637–652

Sprent, J.I. and Sprent, P. 1990, *Nitrogen-fixing Organisms: Pure and Applied Aspects,* Chapman and Hall, London, UK

Stribley, D.P. 1989 'Present and future value of mycorrhizal inoculants', in R. Campbell and R.M. MacDonald (eds), *Microbial Inoculation of Crop Plants,* Special publications of the Society for General Microbiology, Vol. 25, IRL Press, Oxford, UK, pp.49–65

Tartaglia, J. and Paoletti, E. 1988, 'Recombinant vaccina virus vaccines', *Tibtech* 6, February: 43–46

Towalski, Z. and Rothman, H. 1986, 'Enzyme technology', in S. Jacobssen, A. Jamison and H. Rothman (eds), *The Biotechnological Challenge,* Cambridge University Press, UK, pp.37–76

UNDP, 1989, 'Plant biotechnology including tissue culture and cell culture', UNDP Programme Advisory Note, United Nations Development Programme, New York, USA

Van Frankenhuyzen, K. 1993, 'The challenge of *Bacillus thuringiensis*', in J.S. Corey, M.J. Baily and S. Higgs (eds), *Bacillus Thuringiensis: An Environmental Biopesticide*, John Wiley and Sons Ltd., Chichester, UK, pp.1–35

Van Latum, E.B.J. and Gerrits, R. 1991, 'Biopesticides in developing countries: Prospects and research priorities', in C. Juma, J. Mugabe and N. Clark (eds), *Biopolicy International Series No. 1*. Nairobi, African Centre for Technology Studies.

Van Rossum, D. 1994. *The groundnut*-Bradyrhizobium *Symbiosis: Symbiotic, Physiological and Molecular Characterisation*, Copyprint 2000, Enschede, Netherlands.

Venkataraman, G.S. and Shanmugasundaram, S. 1992, 'Blue-green algae: A biofertilizer in rice cultivation', in M.M. Rai and L.N. Verma (eds), *National Seminar on Organic Farming*, College of Agriculture, Indore, India, pp.14–22

Verma, L.N. and Dube, J.N. (n.d.) 'Production and use of biofertilisers in India', unpublished paper, Biofertiliser Production Centre, J.N. Agricultural University, Jabalpur, India

Walgate, R. 1990. *Miracle or Menace: Biotechnology and the Third World*, Panos Institute, London, UK

Waugh, R. and Powell, W. 1992, 'Using RAPD markers for crop improvement,' *Tibtech* 10: 186–191

Whitten, M.J. and Oakeshott, J.G. 1991, 'Opportunities for modern biotechnology in control of insect pests and weeds, with special reference to developing countries', FAO Plant Protection Bulletin 39, pp.155–181

Whitton, B.A. and Roger, P.A. 1989 'Use of blue-green algae and *Azolla* in rice culture', in R. Campell and R.M. MacDonald (eds), *Microbial Inoculation of Crop Plants*, Special publication of the Society for General Microbiology, Vol. 25, IRL Press, Oxford, UK, pp.89–100

Williams, G. 1988, 'Novel antibody reagents: Production and potential,' *Tibtech* 6, February, pp.36–42

Xu, K.P., Greve, T., Callesen, H. and Hyffel, P. 1987, 'Pregnancy resulting from cattle oocytes matured and fertilised *in vitro*', *Journal of Reproduction and Fertility* 81: 501–504

# 8 The Socio-political Context

*Theo van de Sande, Guido Ruivenkamp and Stéphane Malo*

## Introduction

Biotechnologies are neither universal in their applicability nor neutral in their effects. They reflect the interplay between scientific and indigenous knowledge, agro-ecological conditions and socio-political choices.

As we saw in Chapter 6, the biotechnologies used in agriculture are developed through two different innovative systems: the informal system at farm or village level, responsible for most traditional products and processes, and the formal institutional system, linking farmers with research and technology transfer services in the development and dissemination of modern technologies.

The distinctive characteristic of the informal system is that producers themselves are the major innovators. In a process of trial-and-error, they develop solutions to locally perceived problems with the means at their disposal, the use of external inputs being rare. The many examples of indigenous biotechnologies given in previous chapters testify to the strength of this system in developing and disseminating traditional food and medicinal products.

The industrial revolution in Europe during the nineteenth century marked the beginning of the formal institutional innovative system. Technological change was no longer left to small-scale producers. The separation of formal research on the one hand from productive application on the other came into being as a way of mastering, rather than working with, the environment. Scientists became the engineers of new crop and animal varieties, new pesticides and medicines, and of the new and very different farming and food processing practices that accompanied these modern technologies.

Until recently, the most striking example of the industrial approach to agricultural production was the Green Revolution. Instead of applying their tailor-made solutions to local problems, farmers were provided with standard technologies developed by remote institutes. Farmers had to be taught to apply these technologies, to read labels and to follow instructions, especially when using agro-chemicals. Extension became the channel for informing farmers and persuading them to apply innovations.

Despite the Green Revolution and associated attempts to re-organise agriculture in developing countries, many of the world's people still rely heavily on traditional knowledge of plants, animals, insects, microbes and farming systems for their food and/or medicines. The standard solutions of the Green Revolution were often not appropriate to the complex and diverse circumstances of agriculture, especially in the more marginal rainfed areas. Although the exact percentage is difficult to determine, it has been estimated that up to 80% of the world's population still depends on indigenous knowledge for its medical needs, while at least half rely on indigenous knowledge and crops for its food supplies (RAFI, 1994).

Modern science-based biotechnology has provided a new impetus to formal agricultural research. This new set of tools has already demonstrated its capacity to generate innovations for industrial agriculture. Many of the tools, especially those based on recombinant DNA technology, were designed for that purpose. At the same time, science-based biotechnology also holds great promise for improving traditional farming systems in developing countries. As the potential for expanding the use of irrigation, fertilisers and pesticides shrinks, improvements in these systems will become critical to enhancing the well-being of millions of poor people.

The degree to which science-based biotechnology research can be directed towards meeting the needs of resource-poor farmers in developing countries remains to be determined. Among the many institutions that shape the direction of such research are transnational corporations, international agricultural research centres, the national biotechnology programmes and research systems of developing and developed countries, universities, farmers and NGOs. The wide variety of actors involved ensures that diverse biotechnological development paths are followed. Nevertheless, some of these institutions, especially transnational corporations, have much more political influence and economic power than others, particularly farmers in developing countries. Developments in biotechnology often reflect the interests of these more powerful groups (Ruivenkamp, 1992).

This chapter examines the interaction between modern science-based biotechnology and traditional farming, with special reference to the interests of the rural poor in developing countries. It also discusses the controversial issue of intellectual property rights and the compensation of farmers for their role as the creators and custodians of the genetic diversity on which science-based biotechnology depends.

## The agro-industrial chain

Agriculture can no longer be regarded as a sector made up of independent farmers or farming communities. In the developed countries especially, most farmers are now integrated in the global economy and must be considered as but one link in the agro-industrial chain of production. This chain consists of:

- producers of inputs, such as seeds and agro-chemicals,
- farmers, as producers of agricultural raw materials,
- the food industry, which processes those materials,
- distributors of processed products to the consumer.

Technological advances in the first and third links of the chain have considerable impact on socio-economic developments in the farming sector (Ruivenkamp, 1989). These advances originate either in private enterprises or in public-sector research.

Transnational corporations and specialised biotechnology firms in the private sector are the prime movers in the design, development and deployment of science-based biotechnologies. They are interested in increasing their turnover and profits by selling new biotechnology products, and in extending their grip on world-wide food production. For the development of new varieties of crops or processes with less commercial potential, public-sector research is and will remain the major source of innovation.

Although the new products and processes so far developed through biotechnology research still fall well short of what is theoretically possible, they are likely to trigger further integration within the agro-industrial chain of production.

## The supply of inputs

The direct impact of the input supply industry has so far been low in most developing countries. In some cases there is a virtual absence of external inputs; in others they are used only on certain cash crops. The changes to existing management practices that accompany new inputs are often not well suited to the farmer's needs. Many inputs are more likely to be adopted by specialised or large-scale farmers than by the typical smallholder, who has less cash to invest in what he or she may see as a risky departure from tried and tested approaches. The agro-ecological heterogeneity of many traditional systems is a further factor mitigating against the development of appropriate technological innovations and inputs for such systems. Many rainfed areas have experienced a decline in their share of national and international markets compared with irrigated areas, in which the conditions of production are more stable and uniform.

Neverthess, change is on the way. The delivery of appropriate technology for resource-poor farmers in marginal areas is now recognised as a major challenge facing the formal public-sector research and development system. The development of improved crop varieties for these more difficult environments started later, proceeded more slowly and is prone to more setbacks than technology generation for high-potential areas, but there have been encouraging signs of more rapid progress in certain commodities and countries in recent years. Maize in Zimbabwe, Kenya and the middle belt of Nigeria, cassava on Colombia's north coast, durum wheat and winter chickpea in West Asia, short-duration pigeon-pea in South Asia and drought-tolerant sorghum in southern

Africa are some of the success stories; although in some cases convincing evidence of adoption (and of impact) remains to be provided. The balance of the limited evidence that is available suggests that the recovery of agriculture in the marginal areas inhabited by resource-poor farmers is patchy, slow, and could still be reversed, but it is nonetheless there.

Despite the progress made by the public sector, input-supply companies remain the major developers of biotechnology-based innovations in agriculture, and their aim, as we have seen, is not to develop inputs appropriate to resource-poor farmers but to make a profit. These companies use technology as a way of increasing farmers' dependence on them, for example by engineering new hybrid crop varieties that preclude the production of seed on the farm, or by designing crop varieties for use in conjunction with a specific pesticide. Improved seeds are generally accompanied by a package of recommended inputs and practices without which they will not perform well.

In recent years hybrid varieties have become increasingly important in various commercial food crops, including maize, sunflower, sorghum, sugar beet, cotton, and many vegetables. Many farmers in the USA and Europe, but also in a growing number of developing countries, now rely totally on such varieties. According to van Wijk (1994), sales of hybrids account for nearly 40% of the global commercial seed business, worth about US$ 15 billion.

The most commonly cited advantage of hybrids is their higher yield potential. To be commercially viable, hybrids must usually offer a yield gain of 15–20% over open-pollinated varieties (OPVs). While many hybrids do indeed offer gains of this magnitude, the risk-averse resource-poor farmer generally pursues yield stability across years rather than higher yields per se.

Another advantage of hybrids over OPVs is that the characteristics of the former are always known to the plant breeder. As a result, new desirable characteristics can be more easily bred into hybrid varieties, in which breeders work with two inbred lines instead of with natural populations. In addition, hybrids can be made extremely uniform in plant height and time to maturity. Such characteristics are useful in mechanised agriculture but of less interest to the small-scale farmer using manual family labour. A clear disadvantage of the genetic uniformity of hybrids is that it makes entire crops susceptible to the same pests and diseases, a further factor discouraging adoption by the risk-averse small-scale farmer.

Finally, hybrids have built-in protection against seed multiplication. Unlike the seed of OPVs, that of hybrid plants has lower yields than first-generation seed. Farmers must therefore buy seed every year in order to continue obtaining high yields. If the breeder keeps the parental line in-house, no competitor or farmer will be able to reproduce the hybrid. The market potential of hybrids is one of the most important incentives for the development of a private-sector seed industry in developing countries. The hybrid makes farmers more dependent on external seed deliveries and on generating an income: they must sell their surplus production on the market to be able to buy new seed for the next season (van Wijk, 1994). Most hybrids have been developed by the private sector, but the public sector has also conducted some research in this area, notably on wheat and maize. This research may have been motivated as much by pressure from the seed industry as by the desire to achieve impact through more effective dissemination.

One new biotechnology that could reduce farmers' dependence on hybrids is apomixis: the production of seeds without fertilisation. Many wild plants are naturally apomictic (for example the common dandelion, *Taraxacum* sp.), but not the domesticated crop species. The introduction of this trait to crop plants would allow them to produce seeds with identical genetic properties, fixing the plants' genetic make-up. Apomixis could be applied to obtain seeds from a hybrid plant while retaining the vigour of the first generation. Research on apomixis in crops such as pearl millet, rice, maize and forage grasses is currently under way at various international agricultural research centres (IARCs). According to Yves Savidan, coordinator of the International Network on Apomixis Research (APONET), apomixis provides small farmers with a tool for removing the hybrid's biological protection against propagation. It could thus undermine the profitability of the private seed industry and permit many more of the world's farmers to use hybrids. The introduction of this trait into crop plants would restore farmers to their traditional role as innovators (Jefferson, 1994).

Tissue culture techniques are another biotechnology that can be used to benefit small farmers. In Nepal (Manandar, 1992) and Vietnam (Nguyen van Uyen, 1991) these relatively simple techniques are used to produce virus-free, high-quality potato plants at prices that are attractive to small farmers. In Thailand, they are being applied to multiply promising cultivars of orchids and other ornamental flowers. Through selection and cross-breeding, small-scale growers try to develop an orchid variety with distinctive characteristics (colour or shape). In the Bangkok Flower Centre, these plants are multiplied and returned to the grower, who is then able to raise the plants and sell the flowers. The Centre is sometimes paid in kind. It keeps some of the plants, raises them and acquires its own share of the market (Uthai Charanasri, personal communication).

Through the development and application of such techniques as tissue culture, cell fusion, rDNA markers and gene transfer, a number of input supply companies are attempting to alter and control the genetic structure of the seeds they sell. These techniques make it possible to change the genetic information that affects how the plant grows and responds to its environment. They make agriculture less dependent on a given natural environment, since yields are determined less by specific conditions such as soil characteristics and rainfall and more by the scientific and technological knowledge that is embedded in the new inputs, whether seeds, fertilisers, pest control methods, or veterinary drugs and vaccines. If they become widely used, these inputs, and the companies that deliver them, will increasingly determine what farmers produce, where, when and how.

Science-based biotechnology is also applied to improve the health and productivity of livestock. The American Society of Tropical Veterinary Medicine has identified many potential benefits of biotechnology for livestock, including improvements in reproductive efficiency, animal nutrition, and the diagnosis of viral, bacterial and parasitic diseases. The genetic engineering of embryos could assist in bringing about rapid gains in animal productivity. Also of major potential are several vaccines, diagnostic kits, and various treatments to improve the well-being and health of nomadic herds. In all of these cases, science-based biotechnology offers solutions that are not available through traditional means (Yilma, 1992).

As early as 1962, Rachel Carson's seminal work, *Silent Spring*, cast serious doubts on the wisdom of using pesticides in agriculture. Her book highlighted the toxic effects of pesticides on farmers, consumers and the environment, leading to health problems and to the disruption of ecosystems (Dinham, 1993). For many small-scale farmers, pesticides are either unavailable due to logistical problems or else too expensive. Their limited cash resources are more likely to be spent on the seed of new high-yielding crop varieties. Yet if farmers adopt such varieties without using the prescribed pesticides and fertilisers, they risk achieving lower yields than if they had continued using their traditional variety, with its resistance/tolerance to local pests and diseases.

Several strategies have emerged to deal with these problems. The Food and Agriculture Organisation (FAO) has formulated the International Code of Conduct on the Distribution and Use of Pesticides (FAO, 1990), while biofertilisers and biopesticides have been developed to replace their chemical counterparts.

The Code of Conduct, adopted by the FAO conference in 1985, is a voluntary one. It was formulated in response to concern about the supply of pesticides to countries with insufficient mechanisms for registering pesticides and ensuring their safe and effective use. A further concern was that the residues of pesticides not needed or permitted in particular countries might nevertheless enter those countries in agricultural products imported from other countries. According to the Code, every effort should be made to apply pesticides only in accordance with good and recognised practices.

The export to some developing countries of pesticides banned or severely restricted in other countries has been the subject of public concern. In addressing this issue, the FAO conference in 1989 agreed to introduce provisions for Prior Informed Consent.

Exporters of banned or restricted substances should ensure that the importing country is properly informed.

As acknowledged in its Preface, the Code of Conduct will not solve all the problems associated with the development, distribution and use of pesticides. Nevertheless, it should go a long way towards defining and clarifying the responsibilities of the various parties involved. It is of particular value in countries which do not yet have control procedures.

The corporations producing pesticides are facing hard times. The economic recession, rising research costs and restrictions on the use of their products combined to reduce their turnover and profits by 10% to 20% in 1992 (Heselmans, 1994). Mergers and take-overs are likely to become the norm in this beleaguered sector. In their struggle for survival, many companies are turning to science-based biotechnology for answers. Their search for new and more powerful pesticides is combined with the development of genetically engineered crops that are resistant to the new pesticide. A company can greatly improve its market share if it can design a package consisting of an effective pesticide and a crop that can survive the pesticide. For example, research is under way to produce maize varieties that are resistant to Basta, produced by Hoechst Schering Agrevo, or Roundup, produced by Monsanto. Basta-resistant maize is expected to be released in Canada some time in 1995 (Heselmans, 1994).

Research and development in biopesticides has made some progress over the last decade or so, but still lags way behind that on conventional pesticides. A few applications, such as the biological control of cassava mealybug developed by the International Institute of Tropical Agriculture (IITA), appear to have brought widespread benefits to resource-poor farmers, avoiding serious crop losses (see Box 8.1). In addition such technologies bring environmental gains and savings in foreign exchange, as conventional pesticides no longer have to be imported and applied. In other cases, biopesticides developed

---

**Box 8.1**
**Mealybug meets wasp**

In the early 1970s the cassava mealybug, a pest native to South America, reached Africa. Having no natural enemies to contend with, it spread rapidly, invading 32 countries and causing crop losses of up to 80 per cent.

Scientists at CIAT and IITA decided to investigate the pest in its South American habitat. They discovered that a parasitic wasp, *Epidinocarsis lopezi*, could be used to reduce the pest population to below damaging levels. IITA researchers brought the wasp to Africa, multiplied it and released it from the air at 160 locations in infested areas.

By the late 1980s cassava production in Africa had risen by 10.2 million tonnes a year. It was estimated that for every one US dollar spent by the project, African farmers reaped US$ 149 in increased crop productivity. In addition, the use of expensive and potentially harmful chemicals had been avoided.

This work clearly demonstrates the value of a public-sector international research effort dedicated to protecting the environment and alleviating poverty, goals which the private sector does not address.

Source: CGIAR (1996).

through indigenous knowledge are unfairly appropriated by the private sector (see Box 8.2, p. 192).

## The processing of outputs

For the food processing companies, dependence on a specific crop to produce a specific food is a limitation, reducing their ability to cut costs by making more flexible use of a broader range of cheaper raw materials. These companies have developed and applied the tools to break this connection, allowing one ingredient to be substituted for another. In this way crops become raw materials whose component parts can be broken down and re-assembled in food packages. The agricultural sector becomes a sub-sector of the food processing industry.

This trend is strongly associated with the development and introduction of science-based biotechnology, which allows intervention at the cellular or molecular level. For example, cocoa butter is no longer necessary for the production of chocolate. Enzyme technology has been developed to transform other fats, including sunflower or soya oil, into a comparable but cheaper product. Currently, the European Union is debating whether or not to allow the use of cocoa butter substitutes in chocolate. Some countries already permit the use of up to five per cent substitutes while still allowing the product to be called chocolate. If this policy is adopted by the Union as a whole, the demand for cocoa butter from developing countries will fall drastically. This is but one of many examples of the indirect effects of biotechnology development on resource-poor farmers in developing countries.

Such technologies completely negate the normal process of food production. No longer does production begin with the harvesting of a specific crop or livestock commodity, for example milk to be converted into butter, but instead with the image of the end product, since the substitute product must be made as acceptable to the consumer as the original one, if not more so. Interchangeable agricultural components (such as vegetable oils of various kinds for the production of margarine) are selected and used to produce new products, which are sold to the consumer as offering a healthier, more nutritious or higher status alternative.

The concept of an agricultural product as merely an aggregate of biochemical components, each of which can be isolated and used as a general input into food processing, will stimulate the development of many new food products. Biotechniques will emerge as the natural tools for use in this new approach, thereby influencing the direction of biotechnological research, which will be strongly market-driven.

New products are marketed with the greatest success in urban areas, where they carry the attractions of modernity and convey enhanced status on their consumers. The glamorous image of such foods stands in marked contrast to that of traditional food-stuffs. The latter are made almost entirely from locally available raw materials which have been combined in recipes developed down the generations. These foods are not necessarily better or worse nutritionally than the more formal, modern foods, but from a social point of view they carry certain major advantages, being generally cheaper, more accessible, more attuned to local tastes and more likely to create income and employment locally.

Thus science-based biotechnology can be used to alienate food production from agricultural production and to further concentrate wealth and power. Clearly, the lives of traditional small-scale food producers and processors in developing countries are likely to be radically affected. The increased supply and interchangeability of agricultural products, while it will provide new opportunities for some, will often lead to increasing competition between farmers, accompanied by declining prices and incomes. Those who stand to lose most are the producers of the traditional export commodities that will be displaced by the alternatives made possible through new technology. In the food processing sector, changing tastes will again create some new opportunities, but the larger the share of the market taken by mass-produced products, the smaller the share left for small-scale food processors.

# Intellectual property rights

Science-based biotechnology derives its great political significance from the fact that research and development concentrate on the strategic links in the agro-industrial chain of production. Inputs such as seeds, fertilisers, pest control systems and drugs, enzymes, amino and fatty acids have a social and political importance that goes far beyond their mere economic value. An international system of intellectual property rights that sanctions attempts to privatise and monopolise their production and marketing would give patent holders enormous power over global food production.

## The value of genetic resources

For centuries, the reproduction of crops was controlled by farmers and food producers in their communities. Seeds for next season's crops were retained from harvests and exchanged between communities. Selection techniques, which were integrated with the annual cropping cycle, were used to adjust varieties to changing agro-ecological conditions and processing requirements (see Chapter 6).

Over the past 200 years, various institutions including private-sector companies, national and international government and much more recently, non-government organisations have evolved which have an interest in collecting, conserving and/or improving these genetic materials. They obtained plants and seeds free-of-charge, principally from peasant farmers and other rural people in the developing countries. The unrecompensed appropriation of these valuable materials has been justified by regarding and eventually defining germplasm as the "common heritage of mankind" (Wilkes, 1983); a public good for which no payment is necessary or appropriate.

The concentration of seeds, cuttings and whole plants in genebanks or botanical gardens, mostly administered by scientists from the North, has proceeded apace since the "golden age of plant hunting" in the late nineteenth century (Klose, 1950; Brockway, 1979). More recently, such agriculturally oriented collections have been supplemented by those of medical research agencies. Seeds stored in genebanks are traditionally made available free-of-charge to any bona fide user (usually other scientists).

Under this system of free exchange, germplasm often finds its way into the hands of professional breeders in the private-sector companies of industrialised countries. In this case it acquires a monetary value which, in most cases, is greater than its value to the resource-poor farmer. Contrary to the farmers, who are predominantly interested in the use value of germplasm, professional private-sector breeders are interested in the exchange value of germplasm; the return on their investment. One of the systems set up to protect the interests of professional breeders is known as the plant breeders' rights (PBRs). For this purpose, a certificate is issued describing the new characteristics the breeder has been able to breed into the variety in a stable way. The certificate acknowledges the work of the breeder and states the name of the new variety. Anyone who wants to multiply the variety specified in the certificate has to be licensed by the holder of the certificate and has to pay a fee. Other breeders, however, are allowed to use the protected material of their competitors free-of-charge as starting material for their own improved varieties. This constitutes what is known as the breeders' exemption. Similarly, farmers are allowed to withhold part of their crop as seed material for the next crop without having to pay additional fees to the breeder. This is known as the farmers' privilege.

The emergence of modern biotechnology companies has put this system under pressure. Although receiving raw materials free-of-charge, the seed and pharmaceutical companies are unwilling to share the results of their research in the same vein. They argue that the concept of value added entitles them to charge all users for their products, and that they are entitled to seek to maintain their share of the market by withholding information and materials from potential competitors.

Genetic resources seem destined to become the "oil of the information age". As the value of genetic materials of all kinds increases, a struggle is developing over the social and legal arrangements in place to regulate their appropriation and exploitation (Kloppenburg, 1988; Fowler and Mooney, 1990; Juma, 1989). The companies and governments of the industrialised countries have sought global extension of a legal framework which would give them formal proprietary rights to the new seeds and drugs they are developing. In response, international organisations and NGOs have proposed that farmers should be compensated for their role in the creation and conservation of natural biodiversity.

## Patents and breeders' rights

Patents confer exclusive rights to their holder by granting a legal monopoly on a novel and useful invention. Patent protection is usually awarded for a limited period, mostly less than 20 years.

The International Convention for the Protection of Industrial Property, held in Paris in 1883, was a major step in the internationalisation of intellectual property protection. The Paris convention launched a debate on the ways and means of protecting agricultural property rights that continues to this day. Some have advocated the extension to agriculture of the patents used to cover industrial innovations. Others have claimed that the procedural requirements for granting patents made it extremely difficult to protect plant varieties and other organisms.

In the United States, the attempts to legalise the granting of patents on new crop varieties dates back to 1930. In that year, lobbying efforts from the seed industry led to the ratification of the Plant Patent Act (PPA), which provided protection for asexually propagated plants. In 1970, it was the turn of plants propagated through pollination to be legally protected, under the Plant Variety Protection Act (PVPA). Both PVPA and PPA allow/aim for the protection of intellectual property rights to new plant varieties. The way in which they protect these rights shows more resemblance to the way in which these rights are protected under PBR, than under patent legislation, despite the name of PPA. Contrary to patent protection, in both cases the breeders' exemption applies.

Ten years later, the US Supreme Court made legal history in the Chakrabarty case by issuing a patent for an oil-degrading bacterium. This decision, which was based on the principle that, anything under the sun that is made by man, is patentable, opened the way for the legal protection of a wide range of organisms and plant parts, as well as whole new plant varieties.

Thus, in 1985, the United States Board of Patents Appeals and Interferences reversed a previous trend and permitted the granting of a patent for a maize variety containing a high level of tryptophan, an amino acid. The ensuing tendency to patent living material reached new heights in 1992, when a Wisconsin-based biotechnology firm, Agracetus, was granted rights over all forms of genetically engineered cotton (Mestel, 1994). This was followed by a similar 'species' patents for transgenic soybean (granted) and rice (pending) (Tribe, 1994). Claiming intellectual property rights for genetic material, genes and many other natural products has now become routine in the United States. Often, such claims rest not only on materials originating in developing countries but also on research conducted there (Box 8.2). The word, biopiracy, has been coined to describe them.

**Box 8.2**
**The case of endod**

In Ethiopia, the berries of the African soapberry *(Phytolacca dodecandra)* are used to make a natural soap. In 1964, an Ethiopian researcher found that, in rivers where the women used *endod* (the indigenous name for the soap) to wash their clothes, the zebra mussel *(Biomphalaria)* seldom occurred. Apparently, *endod* was capable of killing the mussel. Since the zebra mussel can act as a host to the parasite causing the human disease bilharzia, the molluscicidal effect of *endod* was extremely useful to indigenous Ethiopian communities.

*Endod* can be produced without highly sophisticated technologies. The soapberry must be harvested while still unripe, dried in the shade and then ground to a powder. About 300 million people worldwide suffer from bilharzia, so a cheap alternative to synthetic molluscicides seemed very attractive. The only synthetic product available with similar toxicity and degradability characteristics was about 50 times as expensive. However, the chemical industry did not show much interest in producing *endod*. A product that can be produced that easily is not commercially attractive to them, according to one of the researchers involved. Obtaining funding for public-sector research and development also proved difficult: donor agencies such as the World Health Organization (WHO) target the bulk of their funds to research on AIDS and malaria.

In the end, a research group at the University of Toledo, in the USA, picked up the research and began investigating whether *endod* could be used to remove the zebra mussel from the pipes of hydro-electric power plants. The University has now been granted a United States patent on the use of *endod*. To get the patent, it conducted one day of experimentation, then spent four months on legal and scientific work to verify the initial evidence.

Opponents to the claim argue that the real work was done by Ethiopian scientists and, above all, by poor Ethiopian communities.

Sources: RAFI (1994); Smit (1994).

With the exception of micro-organisms, which are considered patentable, the European countries do not (yet) mirror the United States in their propensity to patent living material. Article 53(b) of the European Patent Convention establishes the exclusion from patent protection of plant or animal varieties and essential biological processes for the production of plants or animals. The European legislators nevertheless issued a plant patent (the exception that proves the rule) to the US firm Lubrizol in 1989. The patent reportedly covers any transgenic plant produced with a gene under the control of a plant promoter, using the agrobacterium T-DNA for gene transfer (Abbott, 1992). The decision was justified on the grounds that the plant in question was not considered to be a variety, which is defined by the Technical Board of Appeal of the European Patent Office as "a multiplicity of plants which are largely homogenous in their characteristics, that remain stable after every propagation" (van Wijk et al., 1993).

About half the signatories to the Paris Convention are developing countries. Most of them have in the past excluded novel plants, animals and micro-organisms from protection. However, legislation recently enacted in South Korea allows plants of any kind to be patented. In addition, South Korea, Israel, Jordan, the Philippines and Zimbabwe have accepted the principle of patenting genetically engineered micro-organisms if they fulfil the usual patent requirements (Evenson, 1990; Persley, 1990).

While farmers are allowed to replant seeds from a PBR protected variety and breeders to use these varieties, privileges and

*The granting of patents covering all genetically engineered varieties of a species...puts in the hands of a single investor the possibility to control what we grow on our farms and in our gardens. At a stroke of a pen the research of countless farmers and scientists has...been negated in a single act of economic hi-jack.*
Dr G. Hawtin, Director General, International Plant Genetic Resources Institute (cited in Tribe, 1994).

exemptions are far less generous under patent protection than under the PBR system. Whereas in the latter farmers are allowed to replant the seeds of protected varieties, only the breeders' exemption is allowed under patent law. Under patent law, the only exception to the protection is the researchers exemption. Researchers are free to use any protected material, provided that they do not use it for a commercial purpose. In Europe, research on patented material can be performed without permission. In the United States permission has to be asked, but will be granted. Although farmers' privilege is no longer applicable in the case of patented varieties, in practice it is still allowed in Europe.

As an alternative to patent legislation, many countries have chosen to reward innovation in germplasm enhancement by issuing plant breeders' certificates. These countries have ratified the International Convention for the Protection of New Plant Varieties (UPOV). First signed in 1961 by five nations and entering into force in 1968, UPOV had 24 members by the end of 1993, most of them European but also including the United States and Japan. Uruguay and Argentina were the first two developing countries to sign. Chile became the 30th member in December 1995. Colombia and Mexico are expected to follow their example in 1996.

The new version of UPOV, agreed upon during an international conference in 1991, considerably restricts the provision for breeders' exemption. The holder of rights in a variety is granted control over the marketing not only of that variety but also of essentially derived varieties. The most important consequence of this amendment is that a UPOV-protected variety, into which, for example, a new disease, or frost, resistant gene has been inserted by another breeder, can no longer be marketed without the permission of the original certificate holder (Van Wijk et al., 1993).

A second important difference in the new version of UPOV is the restriction of the farmers' privilege. Member states of UPOV wishing to do so may now forbid the sales of the seed saved on-farm. Of course, few if any farmers in developing countries are expected to comply voluntarily with this directive. In most cases, these countries have no patent legislation anyway, with the result that the violation of property rights is impossible there. Even if backed by fully fledged patent legislation, few governments in developing countries would be either willing or able to enact and implement such laws or to check whether they are being obeyed. The bureaucracy and the infrastructure necessary for that degree of control over the lives of small farmers in rural areas is wholly lacking.

The governments of several industrialised countries, as well as some international organisations, are putting increasing pressure on the governments of developing countries to introduce and enforce intellectual property rights. Leading this movement are the United States and the General Agreement on Tariffs and Trade (GATT). The issue of property rights as they affect international trade was one of the most extensively debated issues in the recently completed Uruguay Round of GATT (GRAIN, 1993).

In countries serious in their desire to take advantage of science-based biotechnology, the introduction of intellectual property rights appears to be a prerequisite. Increasingly, foreign technologies will not be made available unless legal protection is provided. Developing countries using imported protected materials in their breeding programmes are exposing themselves to the risk of trade sanctions and, perhaps more seriously, the threat of being removed from GATT's Generalized System of Preferences (GSP) (Acharya, 1991). Following the passing of the US Omnibus Trade Act by the Reagan administration in 1988, the United States has threatened to retaliate with trade sanctions against several countries, among them Brazil, China and Thailand. In response, most of these countries have now undertaken legal reforms to introduce adequate protection (Hobbelink, 1991; Acharya, 1991; van Wijk, 1992).

## Biodiversity and farmers' rights

The increasing pressure to limit the farmers' privilege and the breeders' exemption, coupled with the lack of rewards for the informal innovative efforts of small-scale farmers, have prompted several initiatives to create a more equitable framework for dealing with indigenous crop germplasm. Essentially, what is proposed is a quid pro quo: if farmers in developing countries have to pay royalties for patented seeds, then the producers of these seeds should no longer be allowed to exploit the "common heritage of mankind" free-of-charge, but should recompense farmers as the creators and custodians of that heritage.

The Biodiversity Convention, concluded in Rio de Janeiro in 1993, aspires to be a global mechanism of the kind required to reward communities in this way. However, according to some critics Rio was more a biological GATT than anything else; a debate over how the earth's resources should be exploited rather than how they should be protected. The Convention sanctioned the global hegemony of the existing legal framework for the appropriation and patenting of genetic material. It did not deal with materials already appropriated and stored in genebanks. Nor did it really face up to the issue "farmers rights" to the genetic resources they produce and reproduce (GRAIN, 1994).

In 1983, FAO established a Global System for the Conservation and Utilisation of Plant Genetic Resources as a response to the increasing need to establish a mechanism to coordinate intergovernmental action regarding the safety of plant genetic resources, the ownership of collections, laws restricting the availability of germplasm and intellectual property rights over new varieties. The Global System included a non-binding legal framework (the International Undertaking on Plant Genetic Resources) and an intergovernmental forum (the Commission on PGR). The objective of the International Undertaking is to ensure that plant genetic resources, especially species of present or future economic or social importance, are explored, collected, conserved, evaluated, utilised and made available without restriction for plant breeding and other scientific purposes. The Undertaking is based on the principle that PGR is the common heritage of mankind and, therefore, should be conserved for future generations. This principle has been qualified in a resolution that was adopted by the FAO conference and that was annexed to the Undertaking in 1989. The resolution recognised that the owners of germplasm, through the concept of Farmers' Rights, should be compensated for their contributions to the enhancement of PGR.

There are several practical problems in compensating farming communities, the biggest of which is knowing whom to compensate. Usually, no specific individual, ethnic group or community can be associated with the use or conservation of a specific plant. The plant may be in widespread use across several continents. At the very least it will be used regionally, by several villages or towns and it may have different uses in different communities. In addition, rural communities rarely have the expertise or power to negotiate a fair exchange (Kloppenburg and Gonzales 1994).

The neem tree (*Azadirachta indica*) provides an example of these problems (Kloppenburg and Balick, 1994). Native to South Asia, its useful properties have long been understood by peasant farmers in that region. These properties have also been recognised by companies that operate globally. One of these, the agro-pharmaceutical transnational W.R. Grace, has taken out a patent on the synthesised extract of neem, which it is now marketing as a biopesticide (Burrows, 1993). So far, the company has not offered any reciprocal benefits to the region in which the tree originated. Even if it had, it is not clear whom should be compensated: some Indian farmers, all Indian farmers, NGOs claiming to represent Indian farmers, the Indian government? Additionally, the neem now grows in Africa as well. Should farmers there be compensated too?

Despite these problems, experiments in compensation on a bilateral basis are under way in a number of countries. Box 8.3 provides an example.

*Since we highly value our traditional technologies and believe that (they) can make important contributions to humanity, including 'developed' countries, we demand guaranteed rights to our intellectual property, and control over the development and manipulation of this knowledge.*
Article 44 of the Charter of the Indigenous Tribal Peoples of the Tropical Forests, 1992.

## Conclusion

In the intellectual property rights controversy, new approaches are being put forward every few months. At both national and international levels, the current state of uncertainty with regard to both the legal and institutional arrangements to safeguard the interests of the various parties involved is likely to continue for some time to come.

At least with respect to small farmers producing for local markets, the introduction of legal and institutional reforms is hampered by severe practical problems. It seems unlikely that either the companies or governments in developing countries will be

> **Box 8.3**
> **The INBio-Merck agreement in Costa Rica**
>
> The National Biodiversity Institute (INBio) of Costa Rica is a non-profit scientific organisation created by a national NGO in 1989. Its purpose is twofold: to inventory the biological resources extant in Costa Rica, and to protect those resources by making them available for productive purposes. By establishing their usefulness, the Institute hopes to generate the funds needed for conservation and development. The Institute's central approach is to raise money through the contractual supply of organisms to private companies searching for new sources of drugs, industrial materials and agricultural inputs.
>
> The North American pharmaceutical company Merck & Co. Inc. has expressed interest in the services offered by INBio. Under the terms of an agreement between the two, INBio will provide Merck with 10,000 samples of plants, animals and soils from the country's Talamanca Reserve. Merck will have the exclusive right to study these samples for two years, and will retain the patents to any drugs developed from them. In return, Merck will pay INBio US$ 1 million to fund its activities in the reserve and will donate to the institute an additional US$ 130 000 worth of laboratory equipment. Merck has also promised to pay royalties to INBio for any drug developed from the samples provided. Half these royalties will go to the Costa Rican Ministry of Natural Resources, which will also use these funds to conserve biological diversity.
>
> In setting a precedent for the provision of compensation for the appropriation of genetic resources, the INBio-Merck contract represents an important advance. On the other hand, it establishes a model of compensation that focuses on national rather than on local rights over biodiversity. Although the contract relies heavily on their expertise, the people of the Talamanca Reserve were not consulted at any point in the process, nor are there any provisions to reward them directly, should valuable new products be developed from the genetic resources extracted from their reserve.
>
> Sources: Coughlin (1993) and Kloppenburg and Balick (1994).

able to set up a viable system for monitoring small farmers' compliance with intellectual property rights. The concept of farmers' rights founders on the serious problems of establishing who owns what and of determining who is acting on whose behalf.

The logical flaw behind both approaches is the assumption that germplasm or other inputs are or can be owned by anyone at all. Once this is assumed, then such resources can be thought of as having been expropriated. Then negotiations start on the terms under which the expropriation can be legitimised, and so on.

This narrow and defensive approach is reinforced by the context in which the debate takes place. That context is one in which small-scale food producers and processors apparently run the risk of becoming mere links in the agro-industrial chain of production, dominated by multinationals. These multinationals invest heavily to make farmers increasingly dependent on the external supply of inputs and hence on continuing demand for their produce.

Yet the domination of farmers by private-sector companies is only one possible future for agriculture, and for most peasant farmers and small-scale producers a distant one. That future is by no means inevitable. The direction of science-based biotechnology research and development is, as we have seen, largely determined by power and power relations. If farmers are sufficiently empowered, there are no reasons why they should not be able to influence that direction so that the applications of biotechnology become viable for and accessible to them. Discussions on mechanisms for compensation are a second best option, to be held only when structural solutions have failed.

The next chapter will discuss an approach to research and development that is designed to support farmers' own research efforts by increasing their capacity to experiment at farm or village level and by providing them with better access to, and greater influence over, the resources of the formal biotechnology research and development system.

## References

Abott, A. 1992, 'European Patent Office rejects bid to revoke first plant patent,' *Nature* 357, June 18: 5–25

Acharya, R. 1991, *Intellectual property, biotechnology and trade: The impact of the Uruguay Round on biodiversity*, MERIT, Maastricht, Netherlands

Barton, J. 1991, 'Patenting life,' *Scientific American* 264 (3): 40–46

Brockway, L. 1979, *Science and Colonial Expansion: The Role of the British Royal Botanical Garden*, Academic Press, New York, USA

Burrows, B. 1993, 'Patenting neem: Intellectual property rights or modern piracy?' *Journal of Pesticide Reform* 13 (3): 21

Buttel, F. and Belsky, J. 1987, 'Biotechnology, plant breeding and intellectual property: Social and ethical dimensions,' *Science, Technology and Human Values* 12 (1): 31–49

Carson, R. 1962, *Silent Spring*, Riverside Press, Cambridge, Massachusetts

CGIAR, 1996, *Service through Science*, Consultative Group on International Agricultural Research, Washington DC, USA

Coughlin, M. 1993, 'Using the Merck-INBio agreement to clarify the Convention on Biological Diversity,' *Columbia Journal of Transnational Law* 31: 337–375

Dinham, B. (ed.) 1993, *The Pesticide Hazard: A Global Health and Environmental Audit*, Zed Books, London, UK

Evenson, R. 1990, *Intellectual Property Rights, R&D, Inventions, Technology Purchases, and Piracy in Economic Development: An International Comparative Study*, Science and Technology Lessons for Development Policy, Intermediate Technology Publications, London, UK

FAO, 1990, *International Code of Conduct on the Distribution and Use of Pesticides (amended version)*, Food and Agriculture Organisation, Rome, Italy

FAO, 1993, *Implications of UNCED for the Global System on PGR*, Food and Agriculture Organisation, Rome, Italy

Fowler, C. and Mooney, P. 1990, *Shattering: Food, Politics and the Loss of Genetic Diversity*, University of Arizona Press, Tuscon, USA

GRAIN, 1993, 'GATT, NAFTA and intellectual property rights,' *Seedling* 10 (4): 2–6

GRAIN, 1994, 'Packaging an agricultural biodiversity plan,' *Seedling* 12 (1): 19–22

Heselmans, M. 1994, 'Agrochemie probeert zijn leven te rekken,' *LT Journaal* 17 March: 7–8

Hobbelink, H. 1991. *Biotechnology and the Future of World Agriculture*, Zed Books, London, UK

Jefferson, R. 1994, 'Apomixis: A social revolution for agriculture?' *Biotechnology and Development Monitor* 19, June 1994: 14–17

Juma, C. 1989, *The Gene Hunters: Biotechnology and the Scramble for Seeds*, Zed Books, London, UK

Kloppenburg, J. 1988, *First the Seed: The Political Economy of Plant Biotechnology, 1492–2000,* Cambridge University Press, New York, USA

Kloppenburg, J. and Balick, M. 1994, 'Property rights and genetic resources: A framework for analysis,' in Balick, M. and Laird, S. (eds), *The New Gold Rush: Pharmaceutical Prospecting in the 1990s,* Columbia University Press, New York, USA

Kloppenburg, J. and Gonzales, T. 1994, 'Between state and capital: NGOs as allies of indigenous peoples,' in Greaves, T. (ed.), *Intellectual Property Rights for Indigenous Peoples: A Source Book,* Oklahoma City, Society for Applied Anthropology

Klose, N. 1950, *America's Crop Heritage: The History of Foreign Plant Introduction by the Federal Government,* Iowa State College Press, Ames, USA

Manandar, K. 1992, 'An overview of biotechnology in Nepal,' paper presented at a conference to launch the Asia Network for Small-scale Agriculture Biotechnology (ANSAB), 29 March–1 April 1992, Kathmandu, Nepal

Mestel, R. 1994, 'Rich picking for cotton's pioneers,' *New Scientist,* February 19: 13–14

Nguyen van Uyen, 1991, 'Private small-scale enterprises and development of biotechnology in southern provinces of Vietnam,' paper presented at the International Workshop on Biotechnology for Food Production in Vietnam, Hanoi, 10–12 December 1991

OTA, 1984, *Commercial Biotechnology: An International Analysis,* US Congress, Washington DC, USA

OTA, 1991, *Biotechnology in a Global Economy,* US Congress, Washington DC, USA

Persley, G. 1990, *Beyond Mendel's Garden: Biotechnology in the Service of World Agriculture,* CAB International, Oxford, UK

RAFI, 1994, 'Conserving indigenous knowledge: Integrating two systems of innovation,' Rural Advancement Foundation International, United Nations Development Programme, New York

Ruivenkamp, G. 1989, *The Introduction of Biotechnology in the Agro-Industrial Production Chain,* Jan van Arkel Publishers, Utrecht, Netherlands

Ruivenkamp, G. 1992, 'Biotecnologías 'a la medida': Posibilidades de una evolución centrada en los agricultores,' *Agricultura y Sociedad,* Julio-Septiembre 1992: 83–99

Smit, A. 1994, 'Afrikaans bestrijdingsmiddel krijgt geen kans,' *LT Journaal.* 9 June: 6–7

Tribe, D. 1994, *Feeding and Greening the World,* CAB International, Oxford, UK

Van Wijk, J. 1992, 'GATT and the legal protection of plants in the Third World,' *Biotechnology and Development Monitor* 10, March: 14–15

Van Wijk, J., Cohen, J. and Komen, J. 1993, 'Intellectual property rights for agricultural biotechnology: Options and implications for developing countries,' ISNAR Research Report No.3., The Hague, Netherlands

Van Wijk, J. 1994, 'Hybrids: Bred for superior yields or for control?' *Biotechnology and Development Monitor* 19, June 1994: 3–6

Wilkes, G. 1983, 'The current status of crop germplasm,' *Critical Reviews in the Plant Sciences* 1: 133–181

Yilma, T. 1992, 'The role of biotechnology in tropical diseases,' in Williams, J., Kocan, M. and Gibbs, E. (eds), *Tropical Veterinary Medicine: Current Issues and Perspectives: Annals of the New York Academy of Sciences* 653: 1–5

# Part 3

# Building on Farmers' Practices

# 9 An Integrated Approach to Biotechnology Development

*Joske Bunders, Anne Loeber, Jacqueline E.W. Broerse and Bertus Haverkort*

## Introduction

Generations of farmers in developing countries have attempted to improve their farming and living conditions. Their efforts have resulted in the development of countless valuable (bio)technologies. But it is now generally accepted that the farmer's own innovative capacity can lead to only minor improvements over current practices and technologies. More fundamental change can occur only if local people's knowledge is complemented with that of farmers elsewhere and with formal science-based knowledge. Yet farmers have relatively little access to both these kinds of knowledge (Biggs, 1989; Reijntjes et al., 1992). In addition their opportunities for influencing the agendas of governments and other providers of support remain limited.

If appropriate biotechnologies are to be developed for such farmers, it is essential that the various groups of people involved in the process should interact with one another effectively. These groups include not only farmers and scientists, but also extensionists, NGO staff, representatives of farmers' and women's organisations, policy-makers, donors, traders, processors, input sales staff and others (Bunders and Broerse, 1991).

Many attempts have been made to improve communication between some of these groups, notably between scientists and farmers. However, until recently at least, these attempts have only rarely resulted in an effective research process. Ownership of the final product of research has often remained with the scientists, with the inevitable result that adoption by farmers has been low.

The many people who must interact in the development of appropriate biotechnology have different social, educational and cultural backgrounds, are geographically segregated, and differ in terms of power. They may have different perceptions of the problems of small-scale farmers and of what may constitute appropriate solutions to those problems. One of the major challenges of biotechnology development is to integrate the knowledge of these people without getting ensnared by these differences.

The technology development process thus needs to be carefully structured to ensure good communication between all its participants. The purpose of this chapter is to provide guidelines to that end.

# Participatory Technology Development

Conventional science-based research and development has often been criticised for following a linear process of technology development (Harrison, 1983; Chambers et al., 1989; Shiva, 1991). In this process, universities and research institutes conduct research and develop new technologies, which are then handed out to extension services and might, or might not finally reach end-users.

In response to this criticism, various new approaches have been developed over the past 25 years, starting with farming systems research and evolving into the farmer participatory approaches of recent years (Farrington and Martin, 1987; Scoones and Thompson, 1994). Farmer participatory approaches are collectively referred to as participatory technology development (PTD) (Haverkort et al., 1991; Reijntjes et al., 1992).

*I know of no safe depository of the ultimate powers of the society but the people themselves, and if we think them not enlightened enough to exercise that control with a wholesome discretion, the remedy is not to take it from them, but to inform their discretion.*
Thomas Jefferson, 1820.

## Basic features

In PTD, farmers are encouraged to generate and evaluate indigenous technologies and to choose, test and adapt external technologies on the basis of their own knowledge and value systems (Reijntjes et al., 1992). Hence, PTD builds on indigenous knowledge, combining it with external knowledge and inputs only when farmers themselves perceive the need to do so. It aims at site-specific, culturally adapted and ecologically sound innovations, selected and defined by the farmer (Haverkort, 1992). Besides developing technologies, the PTD process aims to strengthen the farmer's analytic capacity, awareness and self-confidence.

Throughout the process, someone from outside the farming community, for example an extensionist, researcher or fieldworker, supports a group of farmers, acting as a catalyst by adding his/her own knowledge, skills and experiences to those of the farmers. This outsider serves as a researcher or facilitator of the research process, and is based in the rural community (1).

## Phases of the process

No single PTD approach can serve as a universally applicable model. There are many different ways of organising interaction among farmers and of linking farmers' knowledge and experimentation with formal R&D. But despite this diversity, a number of characteristic steps or phases can be distinguished (ETC, 1992; Jiggins and de Zeeuw, 1992). These are:

- induction training of facilitators,
- getting started in the field,

- understanding problems and opportunities,
- looking for things to try,
- organising and conducting experiments,
- implementing, monitoring and evaluating experiments,
- sharing results,
- sustaining and scaling up the PTD process.

In practice, PTD programmes seldom follow all these phases in a linear sequence; some focus mainly on one phase only, others move between phases or repeat them.

## Induction training of facilitators

Before they start their work, facilitators need to be trained in the attitudes necessary for participatory research. These attitudes spring from a feeling of respect for local culture and knowledge, and from an awareness that science-based knowledge and indigenous knowledge are complementary. Since researchers may have been schooled to see "scientific" knowledge as superior and of higher status to "unscientific" farmer knowledge and experimentation, taking a participatory approach implies, in many cases, a reversal of conventional attitudes. For example, one needs to understand that reality can be seen and interpreted in different ways, without there being one single truth or best interpretation (for example, one based on formal, quantitative analysis). Such an understanding will prevent the outsider from rushing into a situation with a predetermined view of the nature of farmers' problems and of the solutions to them. Having a prior technology agenda of their own is one of the criticisms most frequently levelled at researchers in the formal system, even those adopting a farming systems approach to research.

The most important aim of induction training is, then, to make outsiders aware of their own implicit and often subconscious attitudes, which may consist of feelings of superiority, insecurity, or of racial, professional and/or gender biases. Self-awareness is the first and most crucial step in overcoming adverse attitudes like these. A second aim of the training is to impart skills in participatory methods of experimental design, monitoring and assessment of results. Useful techniques which can be used in such methods include semi-structured interviewing, focus group meetings, social map construction, aerial photograph analyses, matrix scoring and preference ranking.

## Getting started

The facilitator undertakes or organises several preparatory activities before beginning in-depth interaction with villagers. The first is to select the villages to be included in the PTD process. This is usually done in response to a specific request. Next, existing secondary data are assembled and analysed, and an organisational inventory is made. The most important preparatory activity is to start building a relationship with the people who will participate in the research process. Basic agreement on how to collaborate has to be reached and the role of the facilitator needs to be explained and accepted.

The outcome of these preparatory activities should include a clear perspective and protocol for joint work, a preliminary understanding of the socio-cultural and agro-ecological circumstances of each participating village, and a core network of individuals and organisations that will play a lead part in the research process.

## Understanding problems and opportunities

Probably the strongest driving force in any PTD programme is the realisation, on the part of farmers, that the programme really addresses their concerns. Farmers' views on the problems they face must therefore be sought, together with their ideas on solutions. Insights into how the local population sees its own situation can be gained by asking questions and holding discussions on trends in the farming system, cause-and-effect relationships, the wider political and socio-economic context, and any resources that could be better used. Information should also be sought on the experiments that local farmers have already carried out, and on their own continuing efforts at local development.

It is vital that this phase should lead to a clear definition of problems and opportunities that is shared by all participants, including the outsider. Several of the techniques mentioned above can be used to achieve this aim.

## Looking for things to try

This phase starts with a more detailed analysis of the problems and opportunities identified in the previous phase. Farmer experts within the village, as well as sources of knowledge from outside, are consulted. The options are critically reviewed (assessing the advantages and disadvantages for different groups) and the criteria for setting priorities are defined. Innovations from within the farming system, as well as those available from formal research and extension, are considered. Farmers systematically screen all ideas and make the final decisions on what needs to be tried first.

Besides a jointly agreed agenda for experimentation, an additional outcome of this phase is improved skills on the part of farmers in diagnosing problems and identifying solutions. There should also be an improved organisational basis for conducting research at village level more systematically.

## Organising and conducting experiments

On the basis of the research agenda, the farmers and the facilitator(s) jointly design experiments that can be managed and evaluated by the farmers. The designs should accommodate farmers' own experimental practices and their criteria for assessing results. Additional farmer-experimenters may be trained if necessary.

This phase should lead to the development of locally applicable technological innovations. It should also yield insights into any likely adoption problems. In addition, greater understanding of the PTD process and improved farmer skills in experimentation may result.

## Sharing results

Farmers learn from each other throughout the PTD process, but it is especially important that they should do so once a successful innovation has been developed. Activities to encourage the sharing of results build on the networks developed during earlier phases. An inventory of existing patterns and channels of farmer-to-farmer communication may serve as the basis for planning activities, which may include organising exchange visits, developing manuals or slide-shows, and training farmer promoters.

At this stage the programme makes preparations for scaling up the PTD process by starting to work with larger groups and seeking to increase the number of villages covered. An inter-village PTD network may develop. A broader range of institutional contacts may also be sought.

## Sustaining and scaling up

The outsider's final objective is to leave the villages with a permanent capacity for implementing an effective PTD process. Training and institutionalisation are therefore major concerns throughout the programme, but particularly during this phase. The emphasis will be on consolidating community networks and organisations, fostering a more supportive external institutional environment and evaluating and documenting the PTD approach.

The outcome of this phase is that the village has a more productive and sustainable farming system and is more self-sufficient in research. The facilitator gradually withdraws from the village and is free to launch the PTD process elsewhere.

## Keys to success

Successful PTD has the following characteristics:

- It is farmer driven. The ideas, priorities and objectives of the farmers determine the course of action throughout the process. Researchers may offer the means for exchanging information or securing external inputs, but should not dominate.
- Outsiders act as facilitators. The outsider has a role in facilitating the development process by adding his/her knowledge, skills and experiences to those of the farmers, by accessing external expertise and resources, by training farmers and by catalysing the creation of groups and networks.

- Feedback and feed-forward mechanisms are created. These are opportunities for the participants to share their ideas and experiences and so to influence the future course of events. Typical mechanisms include village walks, participatory mapping exercises, discussions with key informants, meetings and debates. These occasions allow two important and interrelated requirements for developing appropriate technology to be met: the exchange of information and the reconstruction and verification of participants' opinions. Exchanging information in direct, personal contact in a group setting reveals the diversity of views and the fact that multiple interpretations of a problem or a technological option are possible. Through meetings and debates, the accuracy of the interpretations made by outsiders and other participants can be constantly re-assessed. Before determining a course of action, it is essential to verify its appropriateness by discussing it openly with all the people involved.

- The attitudes of the participants facilitate mutual learning. Perhaps the single most important prerequisite to successful PTD is the willingness to listen. Understanding the actions, motives and goals of fellow participants by listening to one another is the basis for successful cooperation. Connected with the willingness to listen is a feeling of mutual respect. It is important that each person involved in the process recognises the others' expertise and potential contribution to the concerted effort, and behaves accordingly. In the course of PTD, participants are likely to be brought face to face with both each other's and their own biases and insecurities. Working with people from different cultural, racial and/or educational backgrounds is bound to raise problems of this kind. Commitment to a shared goal helps overcome these problems.

Box 9.1 illustrates these characteristics with an example from the Philippines. This example was chosen because it shows clearly how the quality of interaction between farmers and fieldworkers can lead to relevant research.

## The interactive bottom-up approach

When technology development reaches out, as it must often do, beyond the level of the farm or village to the research station or laboratory for inputs, the dynamics of the formal research system need also to be taken into account in research planning and implementation. In addition, developing technology implies dealing with the larger socio-economic system; the worlds of local, national, regional and international policy making and trade. These interactions, absent in early forms of PTD, are especially crucial in biotechnology development, which may require lengthy

**Box 9.1**
**Weed control in the Philippines**

The multidisciplinary research team asked farmers about their current topics of conversation. From among these topics, farmers chose declining soil productivity as the one they would like to discuss further. During visits by the researchers to their fields, the farmers showed them what they meant. In subsequent meetings, the farmers agreed that the problem they wished to address first was the infestation of infertile marginal uplands by cogon grass (*Imperata cylindrica*). The researchers then listed topics for further study, such as why cogon is present and what constraints farmers face in combating the weed.

Surveys provided information on the biophysical and socio-economic factors perceived by farmers as relevant to the cogon problem. At a meeting of all participants, each of these factors was written on a blackboard, in a box, with arrows pointing to the central problem. Key informants explained the relationships in the diagram. The boxes were then redrawn as concentric rings around the problem, the size of each ring

being determined by the proportion of farmers who thought that point important. Another meeting of all respondents was then held to obtain agreement that the diagram indeed reflected what was happening on their farms.

Next, the existing experiments, ideas and knowledge of the farmers were elicited. Most farmers knew that cogon did not germinate in covered soil or grow in shaded areas. Several had observed that it was shaded out or suffocated by vigorous twining plants. Farmers also expressed other ideas for controlling the weed, such as ploughing and planting cassava or sugarcane. To supplement this list, researchers brought up the idea of using herbicides.

At technology screening meetings, key informants and researchers presented these options to farmers' groups for debate. The systems diagram was used to focus the debate, identifying the pros and cons of each option. Farmers judged that ploughing would require too much labour and that draught power and herbicides would be too costly. However, no such constraints were thought to impede the planting of trees or twining legumes. Several farmers wanted to try this option

immediately, but others wanted to see the trees and legumes growing before deciding whether or not to test them. A field trip was arranged for them to see *Leucaena*, *Pueraria* and *Centrosema* species growing at a local research station and on nearby farms.

The farmers decided to test *Pueraria* and *Centrosema*. They expected that the legumes would improve soil fertility as well as shading out the cogon, and that soil covered with low legumes would be easier to cultivate than soil with tall grasses and shrubs. Farmers then chose plot locations and sizes, within limits defined by researchers, who also set the number of replicates and farms. Farmers and researchers together demarcated the plots, and researchers provided the legume seed. Farmers then prepared the land and planted the legumes.

The tasks of data collection and analysis were also shared. Researchers recorded cogon stand densities and took soil samples. They also studied labour requirements. Farmers continually analysed the whole experiment, their keen interest reflecting their involvement in selecting the technology and designing the experiment.

Source: Reijntjes et al. (1992).

laboratory research to obtain the desired results and which raises complex issues such as biosafety, the patenting of germplasm, farmers' rights, and so on.

The interactive bottom-up (IBU) approach to technology development, thus named in deliberate contrast to the top-down approach that has characterised formal sector research in the past, is really an adaptation of the PTD process to fit these more complex circumstances. Experiences with the approach are described in Bunders et al. (1991) and Broerse et al. (1995).

Like farmers, scientists and policy makers respond to suggestions only if they judge them to be rational according to their own criteria for good professional behaviour. It is therefore essential to gain knowledge of these criteria and to integrate them into the process of developing technology, just as we do with farmers' criteria.

In the course of the research process, the circumstances of the participants may change, with implications for both research and development. This applies not only to farmers but also to scientists, policy makers, extensionists and others.

For all these reasons, continuous interaction, with appropriate feed-forward and feedback mechanisms, is essential between all the groups involved in the process (Figure 9.1).

The PTD process emphasises the need to build on and strengthen indigenous knowledge and experimentation (b). Its outcome is recommendations for farm-level experimentation (d) and for improving the approach to technology development (e). The

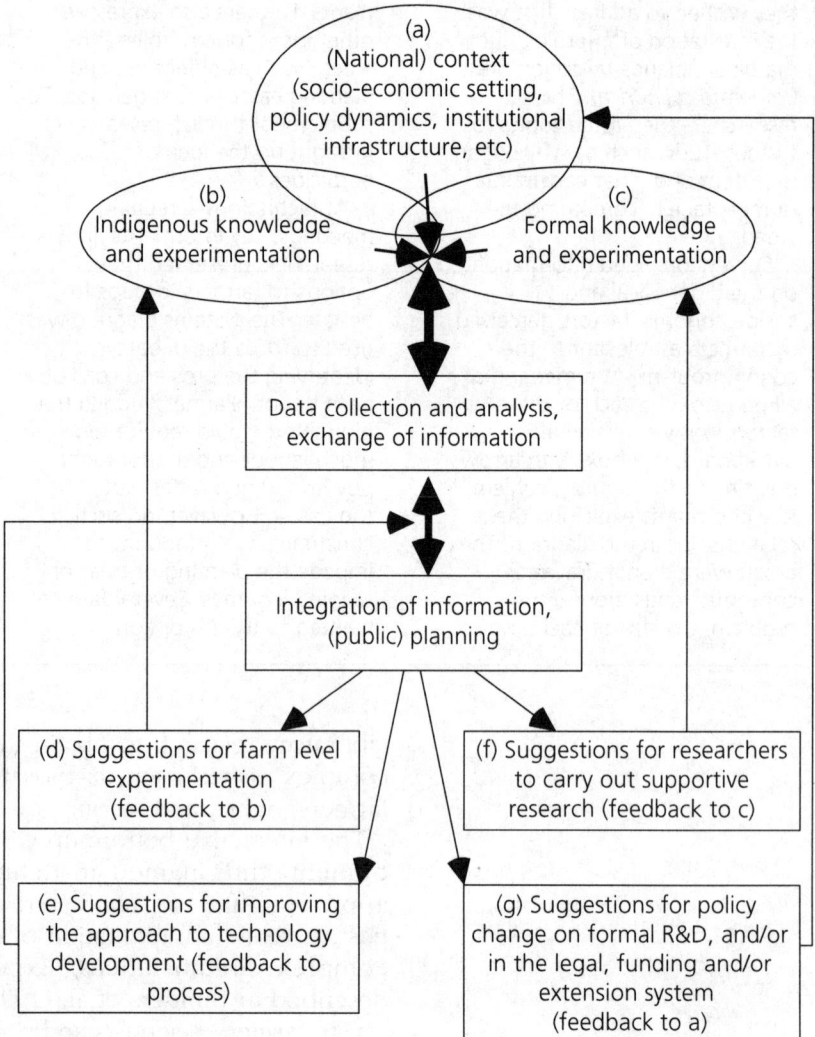

**Figure 9.1**  Feedback and feed-forward between groups in an interactive process of biotechnology development.

process may also yield requests to institution-based researchers (f) and to policy makers (g) for formal research inputs or changes in national policies. However, since the focus of PTD is on the farmer's world and on the dynamics of the farming system, it offers no systematic description of how to influence the worlds of the researcher and policy maker. The PTD process therefore needs to be extrapolated in order to encompass them. This will allow a dynamic exchange of information between all the relevant groups (represented by the bold arrows in the centre of the figure).

## Basic features

In an IBU approach to developing biotechnology, farmers' ideas, priorities and objectives determine the course of action, just as they do in PTD. The potential users of the biotechnology should decide on its design, development and adaptation. All participants therefore need attitudes similar to those required for PTD. However, because a wider circle of actors is involved, the technology development process is planned and implemented somewhat differently.

As described above, an outsider can act as facilitator in the process. However, the facilitator's role involves more tasks, because a wider range of sources of knowledge has to be included. In addition, he/she must also serve as intermediary, helping the different groups involved to bridge the communication gap between them, tackling problems resulting from differences in language or jargon, and identifying and balancing differences in power. The latter include differences in access to information, and in influence over the research and policy agendas.

To accomplish both these roles effectively, a team approach is preferable to having just one person, who would soon find themselves overloaded. The team should consist of members from different professional, scientific and cultural backgrounds. In this way, information from many different sources and disciplines can be collected and integrated.

Opinions on research priorities are likely to differ not only between but also within the different professional communities (for example between men and women, or between social scientists and biologists). Moreover, because of the differences in social, cultural, educational and professional background between these groups, errors in the interpretation of ideas and objectives may easily occur. As a result, it will be difficult to find common ground for action. To reach a consensus, decision making should therefore take place only after organising a live dialogue within and between the groups.

## Phases of the process

The IBU process consists of four main phases:

● initiation and preparation,

● collection, exchange and integration of information,
● public priority setting and planning,
● project formulation and implementation.

These phases can be broadly distinguished on the basis of their chronology, the outcome of the previous phase being the input into the next. Yet the phases overlap, as each consists of activities which may be undertaken several times in a different order throughout the process. The time required for the first three phases is likely to vary from four to ten months.

## Initiation and preparation

The objectives of this phase are to establish and train a multidisciplinary team of intermediaries, and to start work by conducting a literature review and choosing the area for on-farm research. The team members should become familiar with the socio-economic, ecological and political setting.

### Establishing a multidisciplinary team

This is the first activity undertaken once an organisation (for example a group of farmers, a group of extension workers, an NGO, a national government) has decided to launch an interactive biotechnology development process. The team will have the following tasks:

● Collecting, analysing and integrating information (Phase two) for the purposes of public planning (Phase three). This paves the way for an assessment of the appropriateness of biotechnological solutions to the problems perceived by farmers.
● Creating feedback and feedforward mechanisms. These enable all participants to express their point of view and become familiar with that of other participants. The mechanisms consist of interviews, dialogues and informal meetings.
● Organising support for the process. This is done by approaching people who are committed to the objective (generating appropriate biotechnologies for small-scale farmers) and enlisting their support. Once a core of committed people has formed, they can start a dialogue with the doubters.

The team should consist of a small group of permanent members (three to five), and of a fluctuating number of people who, because of their specific expertise, join it for a short period. Its members can be drawn from NGOs, government organisations and/or the formal research community. Balance between these different groups is important: a team consisting only of local people is likely to be unfamiliar with the products and services available from the formal research system; one consisting only of laboratory-based researchers is likely to be too remote from the issues facing farmers. In addition, for the members to act as intermediaries, it is important that they be skilled in inter-cultural and interdisciplinary interaction and rate highly in openness, flexibility, modesty and the ability to listen and respond to others.

*Organising induction training*

The training required is similar to that for PTD. A major aim is to ensure team members have the appropriate attitude for participatory research. A second aim is to acquire skills in participatory field research methods, which are largely qualitative. It is also important that team members should have opportunities for intensive discussion of the approach to developing biotechnology, as well as the implications of their mediating role.

*Getting started*

As in PTD, it is advisable to start with a literature review. This should cover the national economic context and the state of the agricultural sector as a whole, as well as the area under research. It should also provide information on links between the activities of small-scale farmers and other groups. The objective is to get a rough idea of the biotechnological options that are feasible and of the prevailing socio-cultural and agro-ecological conditions.

Together with preliminary discussions with participants, the review helps identify the other relevant actors to be consulted. To develop a feel for the diverging views of these actors, it is important to use as many sources as possible and to collate the views rather than aiming to synthesise them.

The area in which to conduct on-farm research is then chosen. The selection criteria will probably already have been indicated. Typical criteria include representative soil conditions, farm size, market access, age and gender of farmers, class or ethnic group, and so on.

## Collection, exchange and integration of information

In this phase the research needs of farmers are identified. These will be used to establish priorities for further experiments at farm or village level, to set the research agenda of supportive research and development and/or to identify relevant policy issues.

The phase begins with an appraisal of the main characteristics of the agricultural system, from the perspective of each of the parties involved. Next, the status of currently available biotechnology is assessed. In undertaking this activity, the team can generate awareness of the IBU process on the part of relevant scientific institutions, enlisting their support for later activities. This phase allows all involved to obtain a basic understanding of the relevant issues and of the points of view of the various participants.

The team collects data through interviews and dialogues, organises recurrent dialogues with key informants, analyses and integrates this information, and enlists support for the entire endeavour.

*Data collection*

The interviews and dialogues serve four purposes:

- they provide insights into the different perceptions of problems and solutions of different individuals and groups,
- they provide an opportunity to exchange ideas, perceptions and prejudices in an open atmosphere,
- they help bridge the gaps in the often fragmentary information available from the literature review,
- they are a way of identifying people who can act as discussion partners or who are willing to support further initiatives.

By making a round of introductions in the area selected for data collection, the team can list all the individuals and groups that wish to be involved. Through interviews and dialogues, their complementary capacities and resources can be explored. The interviews, which will be conducted both individually and in groups, should include not only farmers but also other people and organisations present in the area. A group discussion with farmers is an excellent way of checking whether information previously obtained from them individually has been misinterpreted. Besides dialogues and interviews, visualised analyses can be used (for example participatory mapping, seasonal calendars, preference ranking, and so on.).

Actors in other communities besides farming are initially identified through the literature review. One way of identifying additional actors at this stage is the "snowball method"; people met in the first round of introductions are asked to name colleagues who might be interested, especially those who might not agree with their point of view. These people are then approached and interviewed. Thus, a wide range of people is invited to participate in the process, reducing the risk that the selection of partners will be influenced by biases and prejudices on the part of the team members. During the interview, the information given by the interviewee is supplemented by the interviewer with information gathered previously, and is discussed with the present discussion partner. Initial feedforward and feedback mechanisms are thus set up. To avoid "steering" by the interviewer, the interview is designed as an open, informal dialogue in which a wide range of topics is covered. These could include the main problems of LEIA, existing or preferred options for solving them, local technological innovations, ideas on potential contributions from science-based biotechnology, and possible side-effects and risks of innovations. The interviewer should introduce the concept of biotechnology only after soliciting the interviewee's opinions on possible solutions to the problems mentioned, so as to prevent the respondent from being preoccupied with the concept of biotechnology in giving her/his ideas.

It is important to note that the activities of collecting and exchanging information are interwoven. The selection of issues raised and the information asked for depends on the information gathered earlier. Throughout the process, the members of the team should reflect critically on their own actions as regards the selection of discussion partners and of information, asking themselves whose ideas and knowledge are solicited, and why some people's analyses and views are recorded while those of others are not.

*Recurrent dialogues*
Some of the people met during the interviews are invited to act as key informants, providing additional information and a means of sifting the information already collected. Their comments, criticisms and support will ensure that ideas that are unlikely to succeed are not pursued further.

Key informants are selected according to their membership of a certain group (for example farmers, researchers, policy-makers or extensionists), their representation of a certain point of view (they should not consist only of proponents but also of opponents, who may provide valuable insights), their analytic capabilities, and their interest in the process and willingness to spend time on it. In all phases, new key informants should constantly be approached, depending on the needs for additional information that emerge. Key informants can be consulted individually or through informal meetings.

*Analysis and integration*
The considerable amount of information available from the literature review, interviews and discussions should result in a general overview of the conditions of LEIA in the area under research and of the potential of biotechnology to bring about improvements. The overview should reflect the differing perceptions of the groups and individuals consulted.

The team then lists the problems identified by the farmers, dividing them into two broad categories, those that can be addressed through biological research and development and those that cannot. Of the problems in the first category, specific attention is paid to those for which a biotechnological solution exists. The feasibility and comparative advantage of this solution over others is assessed. During this exercise, the team regularly obtains feedback from the key informants. Its members should not try to weigh different factors against each other by themselves, because of the likelihood of professional or personal biases.

Box 9.2 provides an example of this critical activity in the IBU process. The project described was funded by the European Parliament through its Scientific and Technological Options Assessment (STOA) programme and executed by the Department of Biology and Society of the Vrije Universiteit Amsterdam. Its objective was to identify ways in which biotechnology could contribute to food security in Zimbabwe.

**Box 9.2**

**Analysis and integration of information on cereal production in Zimbabwe**

After preliminary work in the Netherlands, activities in Zimbabwe began with the establishment of an interdisciplinary team to carry out fieldwork.

Besides a literature review, the team held many interviews. Interviews with farmers were held in two communal areas. The team members made use of various methods to facilitate the exchange of information, including semi-structured interviews, direct observation, focus groups, key informants, preference ranking, historical profiles and trend analyses. The information provided insights into the complexity and diversity of the production constraints facing small-scale farmers, together with their priorities. The group interviews proved extremely helpful in correcting misinterpretations by the team. Over the same period, the team interviewed people employed by organisations responsible for policy making or the development and dissemination of new technologies. Towards the end of the interview, the team members usually confronted the interviewees with other (contrary) ideas and asked them for their opinion on these. The interviewees were also asked to identify other groups or individuals with a stake in the issue under discussion. The resulting information enabled the team to form a view of the conditions of LEIA in the communal areas of Zimbabwe, and on the potential contribution of biotechnology to improving agricultural production in these areas.

The team then listed the problems identified in order of decreasing importance. The one considered most important was poor and declining soil fertility, which applied to all crops. Then came a division between maize and the small grains. For maize, drought susceptibility, limited access to agro-chemicals, lack of appropriate pest control measures, and lack of availability of open-pollinated varieties were the main problems. For small grains, the high labour requirements for production and processing and the lack of institutional support (development of new varieties, milling and marketing facilities, and seed production/distribution) had been cited.

The following areas were identified as ones in which biotechnological options existed: poor and declining soil fertility, drought susceptibility of maize, lack of appropriate control measures for weeds and insect pests in maize, lack of improved varieties of small grains, and inefficient processing and utilisation of small grains.

During the integration phase, diverging views were made as explicit as possible. The differences encountered largely concerned the "supremacy" of maize over small grains in Zimbabwean agriculture, policies and research agendas. Confronted with the opinions of others, most interviewees agreed that these opinions were also valid. But the interviewees drew different conclusions regarding the implications for research. Some argued that the present emphasis on maize did not need to be changed, others that more resources should be devoted to small grains. To stimulate the small grains subsector, some thought that new technologies were needed, whereas others advocated changes in policy. Again, when the underlying rationale for these views was presented, it became clear that some perceptions were not really divergent and that consensus could be reached. Other views were not so easily compatible, but the increased visibility at least led to better understanding of the differences, identifying areas that further research might help to clarify.

Source: Broerse et al. (1995).

*Enlisting support*

In addition to the key informants, the team identifies a wider network of people who are committed to the objectives of the project and may provide support (endorsement, backing, approval, legitimacy) and/or resources (funds, materials, space, staff, time). Developing such a network increases the chances of formulating proposals that can be implemented successfully. This kind of "coalition building" also launches the consolidation process, ensuring the continuation of activities after the project is over.

The farmer [without paper] is describing his analysis of the farming system in the communal areas of Zimbabwe to project staff of the Biotechnology Support Programme to understand the rationale of his actions (*Rommert van den Bos*).

## Public priority setting and planning

Individually or in small groups, farmers or others may decide they wish to test relevant biotechnological options at any stage of the process. At a certain moment, however, it is advisable to organise and hold a priority setting and planning workshop that brings together all the parties involved.

*Organising the workshop*
The objectives of the workshop are:

- to allow review and criticism of the team's findings by a wide audience,
- to legitimise the findings,
- to allow room for new contributions, such as solutions and problems not yet identified by the team,
- to enhance the visibility of the needs of small-scale farmers.

The workshop is thus a means of generating public support for the research process, encouraging implementation of the priorities identified. Discussions at the workshop should lead to consensus on the future direction of biotechnology research and development and on any other relevant matters (links between research at different levels, changes in policy, and so on.). They should also

The farmer is then requested to 'map' his resources
(*Rommert van den Bos*).

lead to clarity as to who will do what, and on the benefits and risks of the interventions planned.

The value attached to the contribution of less influential groups, such as LEIA farmers, should be clearly expressed (for instance, by inviting their representative to make the opening address). At the same time, the presence of highly influential people is also significant, for it signals the importance they attach to the issues facing non-influential groups.

The workshop organisers can take several measures to prevent the participants from misunderstanding each other, such as hiring an interpreter, seating the participants in small groups at round tables, and providing the discussion papers in all relevant languages.

*Holding the workshop*
One of the first events at the workshop will be a presentation describing the outcome of the information analysis and integration exercise. The ensuing discussion allows the participants to

explore different perceptions of the problems and the options for solving them.

Usually there is no initial consensus on priority problems and their solutions. Engineering a confrontation between different points of view may yield consensus if speakers present their arguments well. Any disagreements resulting from merely superficial differences can be sorted out on the spot, while more fundamental differences (policy or ethical issues) can be held over for more thorough discussion in subsequent meetings. If consensus cannot be reached, the parties can at least agree to disagree, and work out a way of handling the differences. Thus, if it does not always yield consensus, the planning exercise at least creates a deeper understanding of the problems.

*Reporting*

After the workshop, the results of the information analysis and integration exercise can be published, together with (or separately from) the workshop proceedings. In this way, participants can see the results of the time they spent. They may also be confronted, in black-and-white, with the perceptions of others, which may differ radically from their own. Diffusing the ideas presented at the workshop may trigger further learning among participants, and among other individuals and groups not so far involved in the planning process. The language used in the proceedings should make the ideas accessible to non-scientists.

The only constraint on spreading information in this way is the illiteracy of some participants. Besides a written report, feedback to farmers can be provided by means of follow-up discussions at village level. These discussions are important for continuing the process or, if it is decided not to continue, for explaining the reasons why.

Box 9.3 provides an example on how this process may take place in a practical situation. The project described is the same one as used in Box 9.2, but at a later phase.

## Project formulation and implementation

The outcome of the previous phase is a list of priorities which could include:

● experiments at farm or village level,
● science-based supportive research on specific topics,
● changes to the policy environment,
● improvements to the approach followed.

This list forms the input to the fourth phase, in which biotechnology projects are formulated and implemented. These projects are, or should be, a direct response to the needs expressed by farmers. As in the previous phases, it is essential that formulation and implementation should be interactive, exploring the options in close collaboration with all concerned.

### Box 9.3
### Priority setting for cereals research in Zimbabwe

The information collected and the options identified in previous phases of the project (see Box 9.2) were presented at a two-day workshop on Biotechnology and Cereal Production by Women in Communal Areas of Zimbabwe, held at the University of Zimbabwe in Harare on 3 and 4 December 1993. Farmers were invited and took their seats beside researchers, extensionists and representatives of NGOs. In addition, several high-profile people, including senior government officials and a representative of the European Parliament, attended.

The first day was devoted to presentations and discussions on agriculture in the communal areas, cereal production and processing, cereals research, and the potential contribution of biotechnology. The results of the team's study were also presented. On the second day, the findings of the study were extensively discussed, and conclusions and recommendations were drafted.

Five women farmers represented the LEIA sector at the workshop. They spoke only Shona and an interpreter was needed. The farmers spoke their minds freely. They wrote down their comments on the first day of the workshop. The plenary session of the second day began by discussing the issues they had raised. Throughout the workshop, other participants regularly asked the farmers for their opinion.

The workshop led to several suggestions for supportive biotechnology research by the formal sector. The topics included biopesticides against pests in maize, genetic engineering of maize to improve its drought and pest tolerance, improved fermentation technologies for processing small grains, development of gene transfer and regeneration protocols for small grains, and genetic mapping in small grains. Suggestions were also made for improving the formal research process, such as applying farmer participatory methodologies, getting research and development workers to take a more holistic approach, and addressing the needs of female farmers.

Several changes in government policy were proposed, namely measures to support farmer participation, the implementation of biosafety guidelines, more emphasis on small grains, increased seed production and distribution, and the training of researchers and extensionists in farmer participatory methodologies. And there were calls for changes in the policies of the European Union, including more donor coordination to avoid overlap and gaps, more attention to small-scale farmers (particularly women), more long-term funding and better feedback mechanisms.

After the workshop the results of the information analysis and integration phase and the proceedings of the workshop were published (Broerse et al., 1994; Gata et al., 1994) and sent to all those interviewed and/or present at the workshop.

Source: Broerse et al. (1995).

Project formulation and implementation can be undertaken at any level and may focus either directly on the farming community or on research institutions or policy making bodies, or any combination of these. Individual farmers or their organisations, women's groups, NGOs, governments, donor agencies, universities, corporations and any others may participate. The incentives for people and organisations to act stem from their involvement in the previous phases. That involvement increases the chances of success.

The projects formulated may differ in time-frame, the actors involved, their financial requirements, their scale, and other aspects. Yet they must be complementary, each contributing to the overall objectives of the IBU process. It may be useful to establish a mechanism, such as a coordination committee, for screening project proposals for their relevance and for any overlaps with other projects. The committee should include members of the team which has mediated and coordinated activities in previous phases.

Women farmer in Hwedza, Zimbabwe, showing her tracking system on the cultivation of peanuts as part of the Biotechnology Support Programme (*Rommert van den Bos*).

### Experiments at farm or village level

The research agenda at local level is set on the basis of the list of priorities. It may be advisable to check whether the biotechnological option that seemed the most promising to the workshop participants is indeed appropriate from the farmers' point of view. Once this has been done, the experiments themselves can be designed and implemented.

Implementation activities are likely to resemble those in PTD. As in the previous phases, the open and free exchange of information is essential.

Experiments should be designed in such a way that farmers themselves are able to monitor and evaluate them. This is of the utmost importance, since the direct support of facilitators will be gradually phased out over the longer term. When the experiments have yielded positive results, the innovations should be disseminated to other farm communities that have shown an interest in the project.

### Science-based supportive research

At the project formulation stage, relevant scientists are invited to discuss the opportunities for supportive research. The institution(s) and scientists are selected according to their commitment to the overall goal of the process as well as for their expertise. With these actors, the coordinating committee assesses the feasibility of the

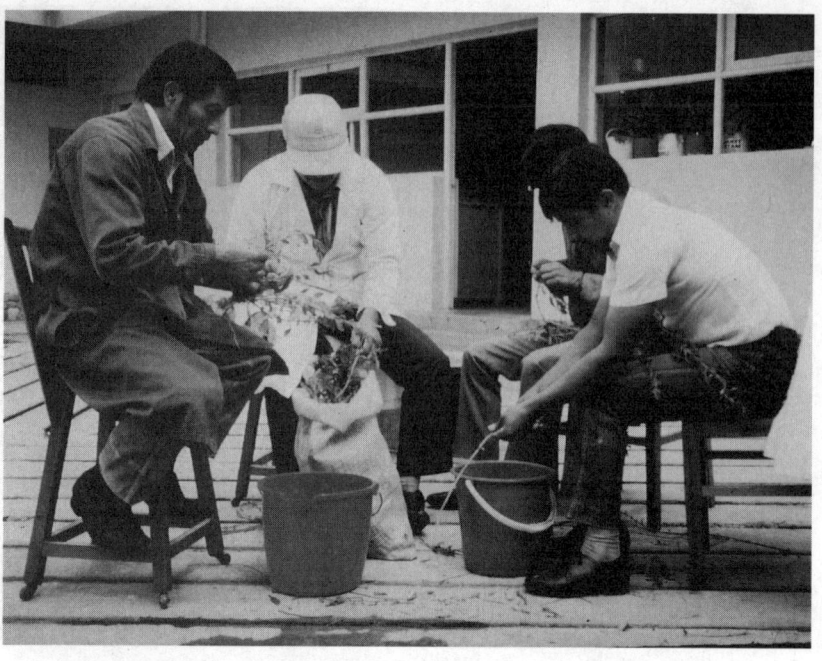

Working together in a participatory research paradigm, villagers and researchers can validate, improve upon and/or find new uses for ancient livestock remedies. Peruvian farmers and scientists prepare and then test a home-made sheep dip that they have compounded from a native wild tobacco, which was traditionally applied as a topical treatment for equine ectoparasites. Both the stockraisers and the researchers agreed that, in its final formulation, this "technology-blended" botanical dip, [invented with inputs from both Western science and ethnoscientific knowledge] proved even more effective in combating ovine ectoparasitism than any of the available commercial products. Such products were too expensive and difficult for farmers to obtain anyway! Ethnoveterinary research and development worldwide has demonstrated that many such folk remedies for livestock diseases with cutaneous signs are highly efficacious. (*Hernando Balazar*).

biotechnological options identified during the public debate. Here especially, the coordination committee must play a mediating role between the researchers and the farming community. Intensive exchange of information between the two groups is essential to ensure that the farmers who are supposed to benefit from the new biotechnology will indeed find it appropriate.

Once agreement on the research agenda is reached, laboratory research can begin. Depending on the biotechnological innovation, this research is likely to take a considerable amount of time (at least two years) before it leads to results that can be tested in farmers' fields. When these results become available, it is prudent to check whether the original problem is still considered relevant and the newly developed biotechnology is still perceived as appropriate. If so, it can be tested and evaluated by farmers and researchers in field trials.

After farmers and scientists prepared the home-made sheep dip together (see picture, page 220), they also test its efficacy together.
(*Hernando Balazar*).

Despite the long time-frame and high costs of such research, efforts must be made to adjust research activities in the laboratory according to any changes in needs expressed by farmers. Likewise, once the products of laboratory research are made available, findings from field trials must be fed back into the laboratory for further product development. Monitoring and evaluation should therefore involve both farmers and laboratory-based researchers, together with a mediator. Efforts must be made to consolidate the links between the research and farming communities, so they continue to cooperate after the project is over.

*Policy changes*
Extension, law, biosafety, education and pricing are all examples of areas in which changes in the policy environment may be necessary or desirable to support the project and enable it to achieve an impact. At the very least, the IBU process should

ensure that the needs of resource-poor farmers start to figure more prominently on policy makers' agendas.

As policy changes can be made only with the backing of relevant policy makers, it is vital to involve these actors at the very start of the process, inviting them to participate both in the exchange of information and in the formulation of projects.

Lobbying and networking by the coordination committee may be important in bringing about the changes identified as necessary.

*Improvements to the approach*

Evaluation and criticism are essential elements of the IBU process, during which it may become evident that adjustments or additional activities are needed. In some cases, evaluation may show that the process could have been more successful if certain issues had been addressed earlier.

Practical experience is the source of most methodological improvements, but reflection on the theoretical assumptions underlying the project may also prove fruitful. It is therefore important to document the outcome of the process in progress reports and articles. Such fora may also be used to discuss methodological issues.

## Sustaining the process

Ideally, technology development should be a continuous process with interactive phases of planning, research, development and adoption/adaptation of innovations. In practice, the record shows that project implementation may be dogged by considerable difficulties. First, project proposals may not be picked up by relevant groups and organisations. These may be interested, but there may be too many constraints and too few incentives for them to actually act. Secondly, proposals may become vulnerable to forces such as standardised working procedures, fixed habits and so on, which may result in their being implemented in a top-down manner or not implemented at all.

To overcome these problems, the current biotechnology research and development system needs to be strengthened. Changes in both individual attitudes and institutional settings are necessary.

## Individual attitudes

Successful biotechnology development depends ultimately on the commitment of the individuals involved, on their flexibility and their willingness to engage in a process of mutual learning. Individual behaviour is often shaped by the professional norms instilled through past training or the current organisational context.

Although the public sector has undergone considerable change in the past 25 years, in many scientific institutions the norms regarding good professional behaviour are still dominated by criteria such as the number of scientific publications and the degree of adherence to reductionist and positivist methods (Pretty and

Chambers, 1993). Personal promotion and institutional recognition may depend on the extent to which these norms are followed.

In addition, efficient implementation of an IBU approach to technology development is still hampered by the difficulties of multidisciplinary research. This is largely because scientists at universities are seldom taught the skills of multidisciplinary and intercultural cooperative research and interaction.

Lastly, it is widely assumed that trained researchers are the true "professionals" or "experts" in the field of agriculture, despite the fact that the literature demonstrates that agricultural systems in rainfed areas have changed more because of the innovative capacities of rural people than because of modern science (Haverkort et al., 1991; Reijntjes et al., 1992).

In the long run, training is the key to bringing about the necessary changes in individual attitudes. Courses in the principles of participatory, interdisciplinary research can be offered to people in organisations such as research institutes and NGOs and included in school and university curricula. In the shorter term, anyone familiar with participatory approaches can play a useful part in inducing change. By inspiring discussion partners to reflect on their own norms and ideas, by networking with people open to this way of working and by inviting institutions to respond to the challenge of changing traditional procedures, people create new opportunities for a more interactive process of biotechnology development. Changes in individual attitudes can produce incremental changes in the institutional setting (Pretty and Chambers, 1993).

## Institutional Settings

For many organisations, participatory or interactive approaches to coordinating and organising technology development are still the exception rather than the rule. These organisations are characterised by centralised hierarchical authority, specialised disciplinary departments, standardised procedures and uniform outputs. Sustaining the IBU biotechnology development process requires that they should become more decentralised, have a flexible organisational structure and produce outputs that respond more to a demand-pull from potential users.

Although institutions can prove remarkably resistant to moves in these directions, change can nevertheless be brought about. The incentive is usually the recognition, within the organisation, that past approaches have failed. This is combined with the commitment of a growing number of staff (usually at management level) to a new, more participatory approach. Often, a funding squeeze is the catalyst for change.

Institutional change can flow naturally from the activities involved in PTD and IBU processes. By identifying and involving committed people and key organisations, by networking, and by establishing feedback and feed-forward mechanisms, the team of intermediaries can help institutionalise improvements. Documenting the methodology followed, together with its results, also helps. Spreading and sharing the experiences, findings and ideas of the project can change the behaviour of actors in related activities.

Change may be frustratingly slow. But one changed institution can act as a catalyst for change in others. This holds true both for local institutions and for national and international ones. More and more organisations, even quite large ones, are now adopting a participatory approach to organising and implementing their activities.

The Dutch Special Programme for Biotechnology and Development Cooperation is attempting to induce change through the research projects it funds in developing countries. The programme has built the principles of interactive biotechnology development into its criteria for funding project proposals (Box 9.4)

Formulating project proposals following guidelines like these can be a good way of promoting change. Yet projects alone are not likely to lead to pervasive change in the way biotechnology is developed. For that to happen, all the efforts and activities geared to adopting a new approach would have to be integrated at national level. An attempt at integration of this kind is the biotechnology programme launched in Kenya, also supported by the Dutch Special Programme on Biotechnology and Development Cooperation (Box 9.5).

---

**Box 9.4**
***The Dutch Special Programme on Biotechnology and Development Cooperation: Criteria for funding project proposals***

Projects submitted for funding are evaluated according to the following criteria:

*Relevance*
- the proposal states the major issues (social, cultural, political, economic or technical) to be addressed,
- it indicates explicitly how the needs of the beneficiaries have been identified and includes a socio-economic analysis of the target group,
- the expected impact of the project in alleviating poverty is assessed,
- the expected impact on women is assessed,
- the expected impact on the environment is assessed,

- the project will contribute to building local research capacity; mechanisms for transferring knowledge are described,
- the proposal describes the risks associated with the research and its results; biosafety issues are addressed,
- the proposal is in line with the country's biotechnology and development policies,
- it indicates the institutional and physical infrastructure required for the dissemination and adoption of new technologies,
- the project has the support of researchers, policy makers, beneficiaries and other stakeholders,

*Scientific aspects*
- the research problem and objectives are clearly stated in the proposal,
- the proposed methodology is sound and takes into

consideration the current state-of-the-art,
- the proposal indicates the scientific qualifications of the individuals and organisations involved,
- opportunities for national and international cooperation are identified,
- the scientific importance of the work and its originality are discussed,

*Operational and management aspects*
- the proposal describes how the project will be managed,
- it describes how continuity will be ensured; it includes a realistic schedule for project implementation,
- it indicates how progress will be monitored and evaluated,
- it lists resource requirements: funds, scientific and support staff, buildings and equipment.

Source: RAWOO (1991).

**Box 9.5**
***The biotechnology
programme in Kenya***

In 1991 a consultant from a local NGO was appointed to coordinate Kenya's previously fragmented efforts in biotechnology research and development. The programme has since undertaken the following activities:

*Studies.* Four inventory and assessment studies were carried out. The first provided an overview of the current status of biotechnology activities in the country and concluded that training, information and documentation, patent issues, biosafety and the role of industry all required more attention.

The second study investigated the current approach to generating biotechnologies. It alleged that most of the developments in biotechnology so far had been supply-driven, and that little had been done to understand the demand side. Support was required in needs assessment, institution building, manpower development, research grant coordination, and research-extension links (Juma and Makau, 1991).

The third study, on NGOs, showed that their scope and coverage is too narrow at present, they often cater only for one technology or work in a small area, but that their activities could be strengthened by collaborating and networking with one another and with government departments. The NGOs studied were thought to be generally responsive to the needs of local people and often applied participatory approaches, so they could be useful in the development and dissemination of biotechnologies.

Lastly, an in-depth case study on indigenous knowledge and practices was carried out in the Kitui District, a semi-arid area with poor infrastructure but good germplasm diversity. One of the findings was that traditional crop varieties had better yield stability and palatability than many improved varieties and were preferred by small-scale farmers for these reasons. The study pointed out that Kenya has no laws to protect its farmers from outside agencies seeking to patent their knowledge. Nor does a system exist for supporting local people in their attempts to improve or commercialise traditional biotechnologies or to apply for patents.

These studies helped identify relevant factors. They also made the key institutions involved in biotechnology development aware of the programme. This helped generate support for later activities.

*Workshops.* Two districts were selected in which to organise farmers' workshops that would identify small-scale farmers' needs. At one of the workshops the participants were a good cross-section of the farming community with regard to age, gender and farm type and size. The opinion of female farmers sometimes differed from that of male farmers. For example, women saw maize as a subsistence crop, whereas men regarded it as a cash crop. These different perceptions led to different recommendations. At the other workshop, larger farmers were over-represented. Nevertheless, all had travelled widely in the district and were familiar with its farming practices and constraints. After lively discussions, the participants agreed on the needs regarding the main crops, livestock and trees. Later, farmers from both districts met again to discuss proposed activities and their involvement in technology development. This time all participants were genuine small-scale farmers and over 50% were women. They confirmed the needs identified during the earlier workshops and expressed their continuing wish to participate.

Next, four workshops, devoted to animal production, crop production, tree production and policy issues respectively, were organised to integrate the information collected. The first three led to the identification of the potential contributions of biotechnology in each of the three fields covered, starting with the needs of the farmers as expressed in the farmers' workshop. The fourth workshop led to recommendations for policy changes in such areas as intellectual property rights and biosafety.

A four day national planning workshop was then held, at which farmers, researchers and policy makers expressed their ideas on priorities. All ideas were subject to debate, following which a plan of action was proposed and discussed. It was agreed to adopt an interactive and participatory approach to the formulation and implementation of a number of biotechnology projects. A national biotechnology forum, consisting of representatives of NGOs, farmers, scientists and policy makers, was set up to facilitate and guide this process. The forum is seeking legal status from government.

*Project formulation.* The forum has identified priorities and invited various institutions to submit projects on them. Its next task is to appraise the project proposals and recommend them for funding.

Source: Joseph Wekundah, ETC, Nairobi, Kenya (personal communication).

# Conclusions

To ensure the development of appropriate biotechnologies, we need both a participatory approach and a broader process of institutional consultation that works from the grass roots upwards, firmly representing farmers' needs to the different worlds of scientists, policy makers and other relevant groups. In this chapter, we have outlined an approach to the development of appropriate and feasible biotechnologies for and with small-scale farmers, the interactive bottom-up approach.

Judging by experiences with the IBU approach thus far, learning can be induced between a wide variety of people from different cultural and professional backgrounds and with diverging views and ideas. Experiences in the creation of feedforward and feed-back mechanisms in the process have been particularly rewarding. In addition, the IBU approach can accelerate the process of change now under way in many formal sector institutions.

The development of appropriate biotechnologies for and with small-scale farmers is feasible. We hope that readers will feel inspired by what has been achieved so far, and will feel stimulated to experiment (more) with interactive and participatory approaches.

## Notes

(1) The term researcher in the PTD literature refers to someone who is actually involved in the innovative activities conducted in farmers' fields over a longer period of time (unlike researchers who are institution-based).

## References

Biggs, S.D. 1989, 'Resource-poor farmer participation in research: A synthesis of experiences from nine national agricultural research systems', special series on the organisation and management of On-Farm Client-Oriented Research (OFCOR), Comparative Study Paper No. 3. International Service for National Agricultural Research (ISNAR), The Hague, Netherlands

Broerse, J.E.W., Bunders, J. F.G. and Loeber, A.M. 1995, 'The interactive bottom-up approach to analysis as a strategy for facilitating the generation of appropriate technology: Experiences in Zimbabwe', in *Industrial and Environmental Crisis Quarterly*, Vol. 9, No. 1, 1955, edited by R. Hoppe and J. Grin, pp.49–76

Bunders, J.F.G. and J.E.W. Broerse (eds). 1991, *Appropriate Biotechnology in Small-Scale Agriculture: How to Reorient Research and Development*, CAB International, Wallingford

Chambers, R., Pacey, A. and Thrupp, L.A. (eds) 1989, *Farmers First: Farmer Innovation and Agricultural Research*, Intermediate Technology Publications, London

ETC. 1992. *Learning for Participatory Technology Development: A Training Guide*. ETC, Leusden.

Farrington, J. and Martin, A. 1987, 'Farmer participatory research: A review of concepts and recent practices', Occasional Paper No. 9. Overseas Development Institute, London

Harrison, P. 1983, *The Third World Tomorrow. A Report from the Battlefront in the War against Poverty,* Second edition, Harmondsworth, Penguin Books, England

Haverkort, B. 1992, 'Cosmology, people's science and indigenous experimentation', paper presented at the Third International Congress on Ethnobiology, Culture and Nature: Directions for Conservation of Diversity, Mexico City

Haverkort, B., van der Kamp, J. and Waters-Bayer, A. (eds) 1991, *Joining Farmers' Experiments: Experiences in Participatory Technology Development,* Intermediate Technology Publications, London

Jiggins, J. and de Zeeuw, H. 1992, 'Participatory technology development in practice: Process and methods', in C. Reijntjes, B. Haverkort and A. Waters-Bayer (eds), *Farming for the Future: An Introduction to Low-external-input and Sustainable Agriculture,* Macmillan Education, London and Basingstoke

Pretty, J. N. and Chambers, R. 1993, 'Towards a learning paradigm: New professionalism and institutions for agriculture', Discussion Paper No. 334. IDS, University of Sussex, Brighton.

RAWOO 1991, 'Criteria for assessing proposals for research in and for developing countries', RAWOO Publication No. 2, RAWOO, The Hague

Reijntjes, C., Haverkort, B. and Waters-Bayer, A. (eds) 1992, *Farming for the Future: An Introduction to Low-external-input and Sustainable Agriculture,* Macmillan Education, London and Basingstoke

Scoones, I. and Thompson, J. (eds) 1994, *Beyond Farmer First: Rural People's Knowledge, Agricultural Research and Extension Practice,* Intermediate Technology Publications, London

Shiva, V, 1991. *The Violence of the Green Revolution: Third World Agriculture, Ecology and Politics,* Zed Books, London and New Jersey

# Afterword

*After receiving your award (US$1000), we used all that money to support a small-scale farmers' extension programme called 'Survival Strategies of Peasants: Amaranth Project'. Our main objective is transferring our research knowledge and materials directly to benefit the peasants. We use hand-outs, publications in magazines, newspapers, radio and television programmes, and are organising field days, food exhibitions, workshops and seminars. Topics include how to grow amaranth foods, rotating and intercropping amaranth, and processing amaranth foods.*
Davidson K. Mwangi, contest prizewinner, from Nanyuki, Kenya, 1995.

This book has revealed the wealth of indigenous biotechnology used by rural people in developing countries. The knowledge possessed by these people, while no longer sufficient to guarantee their future well-being, nevertheless provides the only sound basis for relevant agricultural research and development.

Successfully applying science-based biotechnology to the problems of developing country agriculture is one of the most exciting challenges facing the international research and development community. Of the vast stock of science-based biotechnologies now available, few are as yet directly applicable to the low-external-input production systems of developing countries. But the fact that the new biotechnology is still in its infancy renders the task of steering its future development easier. The new techniques are regarded with high expectations by some, but with caution or even scepticism by others. Proving the pessimists wrong will take imaginative science that is nonetheless firmly linked to users' needs, backed up by considerable investment in training and equipment.

A participatory research process is an essential starting point, for it ensures that users, rather than researchers, determine the research agenda. However, if users are to reap the full benefits, such a process must reach out beyond the immediate circle of farmers and field staff to embrace the policy makers, laboratory-based scientists and others whose decisions and contributions are essential for success.

The editors of this book hope that it makes a useful contribution to the development and adoption of such a process. We would be most interested to receive readers' comments.

# Appendices

TABLE 1 Number of plant species having pest-control properties found in various plant families

| Plant family | Number having pest-control properties | Plant family | Number having pest-control properties |
|---|---|---|---|
| Acanthaceae | 10 | Bromeliaceae | 2 |
| Acraceae | 8 | Bryales (order) | 1 |
| Actinidiaceae | 1 | Burseraceae | 12 |
| Agaricaceae | 4 | Buxaceae | 4 |
| Agavaceae | 6 | Byttneriaceae | 2 |
| Aizoaceae | 5 | Cactaceae | 5 |
| Alangiaceae | 2 | Caesalpiniaceae | 37 |
| Alismataceae | 3 | Campanulaceae | 1 |
| Amaranthaceae | 26 | Canellaceae | 2 |
| Amaryllidaceae | 18 | Cannabaceae | 2 |
| Anacardiaceae | 15 | Cannaceae | 3 |
| Ancistrocladaceae | 1 | Capparaceae | 8 |
| Annonaceae | 14 | Capparidaceae | 1 |
| Apiaceae | 48 | Caprifoliaceae | 10 |
| Apocynaceae | 30 | Caricaceae | 1 |
| Aquifoliaceae | 3 | Caryophyllaceae | 13 |
| Aracaceae | 2 | Celastraceae | 12 |
| Araceae | 30 | Cephalotoxaceae | 2 |
| Araliaceae | 9 | Cephaloziaceae | 4 |
| Arecaceae | 5 | Ceratophyllaceae | 1 |
| Aristolochiaceae | 10 | Characeae | 3 |
| Asclepiadaceae | 31 | Chenopodiaceae | 25 |
| Asteraceae | 261 | Chloranthaceae | 1 |
| Balsaminaceae | 6 | Chrysobalanaceae | 1 |
| Bambucaceae | 1 | Cistaceae | 1 |
| Barringtoniaceae | 2 | Clethraceae | 1 |
| Basellaceae | 1 | Clusiaceae | 8 |
| Begoniaceae | 1 | Cochlospermaceae | 2 |
| Berberiaceae | 1 | Combretaceae | 9 |
| Berberidaceae | 8 | Commelinaceae | 3 |
| Betulaceae | 8 | Connaraceae | 1 |
| Bixaceae | 1 | Convolvulaceae | 18 |
| Bombacaceae | 2 | Corallinaceae | 1 |
| Boraginaceae | 14 | Coriariceae | 1 |
| Brassicaceae | 47 | Cornaceae | 1 |

TABLE 1    Number of plant species having pest-control properties found in various plant families (continued)

| Plant family | Number having pest-control properties | Plant family | Number having pest-control properties |
|---|---|---|---|
| Crassulaceae | 18 | Leeaceae | 1 |
| Crossosomataceae | 1 | Lemnaceae | 1 |
| Cucurbitaceae | 20 | Lentibulariaceae | 1 |
| Cupressaceae | 13 | Liliaceae | 56 |
| Cuscutaceae | 3 | Limiaceae | 1 |
| Cyatheaceae | 1 | Linaceae | 1 |
| Cycadaceae | 1 | Lobeliaceae | 4 |
| Cyperaceae | 13 | Loganiaceae | 9 |
| Cyrillaceae | 1 | Loranthaceae | 2 |
| Depterocarpaceae | 1 | Lycopodiaceae | 5 |
| Dichapetalaceae | 3 | Lythraceae | 5 |
| Dicksoniaceae | 4 | Magnoliaceae | 3 |
| Dictyotalis (order) | 1 | Malpighiaceae | 2 |
| Dilleniaceae | 1 | Malvaceae | 18 |
| Dioscoreaceae | 11 | Marantaceae | 1 |
| Dipsacaceae | 2 | Marattiaceae | 1 |
| Dipterocarpaceae | 2 | Martyniaceae | 1 |
| Ebenaceae | 7 | Meliaceae | 19 |
| Equisetaceae | 2 | Melianthaceae | 2 |
| Ericaceae | 30 | Menispermaceae | 10 |
| Erythroxylaceae | 1 | Mimosaceae | 37 |
| Euphorbiaceae | 83 | Monimiaceae | 1 |
| Fabaceae | 190 | Moraceae | 17 |
| Fagaceae | 11 | Moringaceae | 1 |
| Flacourtiaceae | 10 | Musaceae | 1 |
| Fumariaceae | 1 | Myoporaceae | 1 |
| Gentianaceae | 2 | Myricaceae | 8 |
| Geraniaceae | 10 | Myristicaceae | 3 |
| Gesneriaceae | 1 | Myrsinaceae | 7 |
| Ginkoaceae | 1 | Myrtaceae | 26 |
| Haemodoraceae | 1 | Nostocaceae | 1 |
| Haloragaceae | 1 | Nyctaginaceae | 6 |
| Hamamelidaceae | 3 | Nymphaeaceae | 6 |
| Hippocastanaceae | 3 | Nyssaceae | 2 |
| Hippocrateaceae | 1 | Ochnaceae | 1 |
| Hydrocharitaceae | 2 | Oleaceae | 10 |
| Hydrophyllaceae | 3 | Onagraceae | 3 |
| Hypericaceae | 1 | Orchidaceae | 4 |
| Hypoxidaceae | 1 | Orobanchaceae | 1 |
| Illiciaceae | 1 | Osmundaceae | 1 |
| Iridaceae | 5 | Oxalidaceae | 4 |
| Juglandaceae | 9 | Paeoniaceae | 2 |
| Juncaginaceae | 1 | Pandanaceae | 1 |
| Lamiaceae | 76 | Papaveraceae | 9 |
| Lardizabalaceae | 1 | Passifloraceae | 2 |
| Lauraceae | 15 | Pedaliaceae | 3 |
| Lecythidaceae | 3 | Phrymaceae | 2 |

TABLE 1    Number of plant species having pest-control properties found in various plant families (continued)

| Plant family | Number having pest-control properties | Plant family | Number having pest-control properties |
|---|---|---|---|
| Phytolaccaceae | 5 | Saxifragaceae | 11 |
| Pinaceae | 19 | Scenedesmaceae | 3 |
| Piperaceae | 11 | Schizaeaceae | 3 |
| Pittosporaceae | 1 | Scrophulariaceae | 21 |
| Plantaginaceae | 4 | Selaginellaceae | 1 |
| Plantanaceae | 1 | Selenastraceae | 1 |
| Plumbaginaceae | 4 | Simaroubaceae | 20 |
| Poaceae | 63 | Solanaceae | 70 |
| Podaceae | 2 | Staphyleaceae | 2 |
| Podocarpaceae | 6 | Stemonaceae | 3 |
| Polemoniaceae | 2 | Sterculiaceae | 4 |
| Polygalaceae | 1 | Stryracaceae | 1 |
| Polygonaceae | 24 | Symplocaceae | 3 |
| Polypodiaceae | 41 | Taccaceae | 1 |
| Portulacaceae | 3 | Taxaceae | 4 |
| Portulaceae | 1 | Taxodiaceae | 2 |
| Potamogetonaceae | 5 | Tetragoniaceae | 1 |
| Primulaceae | 6 | Theaceae | 4 |
| Proteaceae | 1 | Theophrastaceae | 3 |
| Punicaceae | 1 | Thymelaeaceae | 12 |
| Pyrolaceae | 1 | Tiliaceae | 4 |
| Ranunculaceae | 75 | Tropaeolaceae | 1 |
| Rhamnaceae | 8 | Typhaceae | 2 |
| Rosaceae | 48 | Udoteaceae | 1 |
| Rubiaceae | 23 | Ulmaceae | 4 |
| Rutaceae | 43 | Urticaceae | 1 |
| Saliacaceae | 7 | Valerianaceae | 3 |
| Salvadoraceae | 1 | Verbenaceae | 35 |
| Salviniaceae | 1 | Violaceae | 4 |
| Santalaceae | 3 | Vitaceae | 11 |
| Sapindaceae | 11 | Winteraceae | 1 |
| Sapotaceae | 9 | Zamiaceae | 1 |
| Sargentodoxaceae | 1 | Zingiberaceae | 13 |
| Sarraceniaceae | 1 | Zygophyllaceae | 4 |
| Saururaceae | 1 | | |
| Total | 2,402 | | |

TABLE 2   Description of some promising pest-control plant species

**Plant characteristics (See legend at end of table for decoding)**

| Scientific name | Common name | A | B | D | H | J | M |
|---|---|---|---|---|---|---|---|
| Aconitum ferox | Indian aconite | 1 | 6 | 3 | 2 | 2,3,11 | 5 |
| Acorus calamus | Sweet flag | 1 | 5,6,9 | 6,10 | 1,2,3,4,5,6 | 2,3,4,26 | 2,5,10,11,13,15,17 |
| Ageratum conyziodes | Goatweed | 3 | 6 | 1 | 1,2,16 | 3,7,10 | 2,5,6,7,8,12,13,27 |
| Aleurites fordii | Tung tree | 1 | 1 | 2,3 | 2,3,4,14 | 3,6 | 4,16,20 |
| Allium cepa | Onion | 1 | 6 | 3 | 4,5,13,14 | 3,4 | 1,3,5,12,21 |
| Allium sativum | Garlic | 1 | 6 | 10 | 1,2,5,13,14,18 | 3,4,6,9,12 | 1,3,5,7 |
| Annona reticulata | Custard apple | 1 | 1 | 1,2 | 1,2,5 | 2,3,4,8 | 1,2,4,5,9 |
| Annona squamosa | Sugar apple | 1 | 1 | 1,2 | 1,2,4 | 3,4,7,8,11 | 1,9 |
| Arachis hypogaea | Peanut | 3 | 6 | 1,2 | 1,6,14,16 | ? | 1,2,6,9,10,12 |
| Artabotrys hexapetalus | Ylang-ylang | 1 | 3 | | 14,18 | 3 | 1,8,17 |
| Azadirachta indica | Neem | 1 | 1 | 1,2 | 1,2,3,4,5,7,14,16 | 2,3,4,7,8 | 2,4,5,6,7,10,11,15 |
| Chrysanthemum cinerariifolium | Pyrethrum | 1 | 6 | 10 | 2,11 | 2,3,4,7,8,10,15 | 13 |
| Croton tiglium | Purging cotton | 1 | 1,2 | 1,2 | 1,2,28 | 3,4,11 | 5,10,12 |
| Datura metel | Angel trumpet | 3 | 6 | 1,2 | 1,2,5,11,21,25 | 4,8,11 | 5,7,8,10,16,17 |
| Datura stramonium | Jimsonweed | 3 | 6 | 3 | 1,2,3,14,16,21 | 3,4,9,12 | 5,7,10 |
| Derris elliptica | Derris | 1 | 2 | 1 | 2,14,16 | 3,4,7 | 5 |
| Haplophyton cimicidum | Cockroach plant | 1 | 2 | 1,2 | 1,2,4 | 2,3,8 | ? |
| Justica adhatoda | Malabar nut tree | 1 | 2 | 1,2 | 1,4,16 | 2,3,4,6,8 | 5,6,7,16,18 |
| Lantana camara | Common lantana | 1 | 2,3 | 1,2 | 1,2,3,4,5 | 3,4,10,12 | 1,2,5,7,8,9,13 |
| Madhuca indica | Mowra | 1 | 1 | 1,2 | 1,2,14,16 | 3,6,9 | 1,4,6,10,12, 15,19,28 |
| Mammea americana | Mamey tree | 1 | 1 | 1,2 | 1,2,4 | 3,8,10,11 | 1,5,9,10 |
| Melia azedarach | Chinaberry | 1 | 1 | 1,2 | 1,2,3,4,5,8,14 | 3,4,8,9,11 | 4,5,7,9,15,19,23 |
| Mundulea suberosa | Sweetcane | 1 | 2 | 1 | 2 | 3,4,7,8,9 | 12 |
| Ocimum sanctum | Holy basil | 3 | 6 | 1 | 1,2,3,5,6,16,16,23 | 2,3,9,10 | 1,2,5,7,10,14 |
| Pachyrhizus erosus | Chinese yam bean | 3 | 4 | 1 | 1,2,4 | 3,4,8,11 | 1,2,3,5,10,11,12 |
| Piper nigrum | Black pepper | 1 | 3 | 1 | 1,2,5,11,14,18 | 2,4,9,10 | 5,9,14 |
| Pogostemon patchouli | Patchouli | 1 | 6 | 1 | 1,2,3,5,28 | 6 | 15,17 |
| Pongamia pinnata | Poonga oil tree | 1 | 1 | 1,2 | 1,2,4,16 | 3,4,8,15 | 2,3,4,5,6,8,9, 10,12,13,15 |
| Quassia amara | W. Indian quassia | 1 | 1,2 | 1 | 2,4 | 3,4,8 | 5,13,21 |
| Ricinus communis | Castor bean | 1 | 2 | 1,4 | 1,2,3,5,14,16 | 3,4,6,9,12 | 5,7,10,15,12, 20,21,22 |
| Ryania speciosa | Not known | 1 | 1 | 1 | 1,2,4 | 3,4,12 | ? |

TABLE 2   Description of some promising pest-control plant species (continued)

## Plant characteristics (See legend at end of table for decoding)

| Scientific name | Common name | A | B | D | H | J | M |
|---|---|---|---|---|---|---|---|
| Schoenocaulon officinale | Sabadilla | 1 | 6 | 1 | 1,2,4,27 | 4,8,11,15 | 5,10 |
| Tagetes erecta | African marigold | 3 | 2,6 | 1,2 | 2,3,14,16 | 1,3,4 | 2,5,7,13,16,27 |
| Tagetes patula | French marigold | 3 | 2,6 | 1,2 | 1,2,5,16 | 9 | 13 |
| Tephrosia virginiana | Devil's shoestring | 1 | 6 | 3,4 | 1,2 | 3 | 2,5,12 |
| Tephrosia vogelii | Vogel tephrosia | 1 | 2 | 1 | 1,2,25 | 3 | 9,10,12,13 |
| Tripterygium forrestii | Three-winged nut | 1 | 3 | 3 | 1,2,4 | 1,2,4 | ? |
| Tripterygium wilfordii | Thunder-god vine | 1 | 3 | 3 | 1,2,4 | 4 | ? |
| Veratrum album | White hellebore | 1 | 6 | 3 | 1,2,4,5 | 3,4,8,11 | 2,5 |
| Veratrum viride | False hellebore | 1 | 6 | 3 | 2 | 4 | ? |
| Vitex negundo | Indian privet | 1 | 1 | 1 | 2,5,14 | 3,4,11 | 5,9,10 |
| Zanthoxylum clava-herculis | So. prickly ash | 1 | 1,2 | 3 | 1,2,5 | 4,10,11 | 4,5,9,17 |
| Zingiber officinale | Ginger | 1 | 6 | 1,2 | 5,14 | 3 | 5,7,14 |

Legend: To avoid confusion, we have used the same codes below as in the *Handbook of Plants with Pest-Control Properties* (Grainge and Ahmed, 1988), from which this table has been derived. However, not all codes have been enumerated in this summarised Table. For example, category C (plant classification) has been excluded. And, within a category, some numbers have been excluded because the above-mentioned plants do not possess that particular attribute. For example, under category *H*, code 12 has been omitted because none of these plants has that pest-control activity (being antitick for animal pests).

**A** *(Plant life cycle)*: 1=perennial, 2=annual.

**B** *(Plant type)*: 1=tree; 2=shrub; 3=woody climber; 4=herbaceous vine; 5=creeper; 6=herb.

**D** *(Habitat)*: 1=tropical; 2=subtropical; 3=temperate; 4=semiarid; 5=Mediterranean; 6=cosmopolitan (1+2+3 categories);

**H** *(Type of pest-control activity)*: 1=anti-insect; 2=insecticidal; 3=growth inhibitor; 4=antifeedant; 5=repellent; 6=attractant; 7=chemosterilant; 8=termite resistant; 9=insectivorous; 10=sticky trap; 11=antimite; 13=antitick; 14=antifungal; 16=antinematode; 18=antibacterial; 20=antibiotic; 21=antiviral; 23=herbicidal; 24=allelopathic; 25=rodenticidal; 26=antifertility (for rodents); 27=antivermin; 28=antisnail/leech; 29=pest-free; 30=synergistic; 31=adjuvant; 32=fish poison; 33=poisonous; 34=anaesthetic/sedative; 35=narcotic.

**J** *(Method of preparation)*: 1=no preparation needed; 2=drying of plant part; 3=aqueous extraction; 4=powdering the plant part; 5=tapping for sap/latex; 6=pressing/distilling for oil; 7=ether extraction; 8=alcohol extraction; 9=ethanol extraction; 10=acetone extraction; 11=petroleum ether extraction; 12=methanol extraction; 13=benzene extraction; 14=ethyl ether extraction; kerosene extraction; 16=chloroform extraction; 17=ethyl acetate extraction.

**M** *(Other economic uses of plant)*: 1=as a food/drink; 2=as animal food substitute; 3=provides fibre; 4=provides materials for making tools; 5=provides medicine/drugs; 6=is a source of fuel/light; 7=is a windbreak; 8=used as a sand binder; 9=for erosion control as a cover crop; 10=as a fertiliser; 11=for soil reclamation; 12=fixes Nitrogen; 13=is an ornamental plant; 14=used as a spice/flavoring; 15=for soap making/as a soap; 16=is a source of dye/ink; 17=is a source of perfume/incense; 18=is a source of honeybee nectar; 19=for wood carving/in carpentry; 20=in paints/varnish; 21=is a source of tannin; 22=is a source of paper; 23=is a source of beads for jewellery; 24=used in weaving; 25=is a source of wood preservative; 26=is a source of rubber; 27=is a source of sulfur; 28=is a source of cooking oil/fat; 29=used in cigarettes.

233

## List of pests controlled by each of the 12 plants listed in Table 2:

### *Acorus calamus*

*Aedes aegypti, Athalia proxima, Aulacophora foveicollis, Bagrada picta, Bombyx mori, Callosobruchus analis, C. chinensis, Ceratitis capitata, Cimex lectularius, Culex fatigans, Dactynotus carthani, Dacus cucurbitae, D. dorsalis, Dermestres maculatus, Dysdercus cingulatus, D. koenigii, Empoasca devastans, Graphosoma italicum, G. mellonella, Latheticus oryzae, Lipaphis erysimi, Musca domestica, M. nebula, Oryctes rhinoceros, Pericallia ricini, Pieris barassicae, Pyrrhocoris apterus, Rhizopertha dominica, Sitophilus oryzae, S. cerealella, Sitrotroga cerealella, Spodoptera litura, Thermobia domestica, Tribolium casteneum* and *Trogoderma.*

### *Allium cepa*

*Altenaria tenuis, Aspergillus niger, Botrytis allii, Callosobruchus analis, Ceratocystis ulmi, Claviceps purpurea, Colletotrichum circinans, C. lindemuthianum, C. trifolii, Corynebacterium michiganese, Curvularia lunata, C. penniseti, Dermacenter marginatus, Diplodia maydis, Drechslera graminea, Fusarium culmorum, F. graminearum, F. moniliforme, F. nivale, F. oxysporum, F. poae, F.* sp., *Gibberella fujikuroi, Haemaphysalis punctata, Helminthosporium* sp., *Ixodes redikorzevi, Myrothecium verricaria, Phyllobius oblongus, Phytodecta fornicata, Pieris napi, P. rapae, Rhipicephalus rossicus, Schistocerca gregaria, Tribolium castenum, Ustilago avenae, Venturia inaequalis* and *Verticillium albo-atrum granarium.*

### *Allium sativum*

*Aedes aegypti, A. nigromaculis, A. sierrensis, A. triseriatus, Agrobacterium tumefaciens, Alternaira tenuis, Aspergillus niger, Botrytis allii, Callosobruchus chinensis, Cephalosporium sacchari, Ceratocystis ulmi, Cercospora cruenta, Cladosporium cucumerinum, C. fulvum, Claviceps purpurea, Colletotrichum capsici, C. circinans, C. lindemuthianum, C. trifolii, Corynebacterium flaccumfaciens, C. michiganese, Culex peus, C. quinquefasciatus, C. tarsalis, Curvularia lunata, C. pennisetti, Dermacentor marginatus, Diplodia maydis, Drechslera graminea, D. oryzae, Dysdercus cingulatus, Erwinia aroideae, E. carotovora, Fusarium culmorum, F. graminearum, F. moniliforme, F. nivale, F. oxysporum, F. poae, F.* sp., *Gibberella fujikuroi, Glomerella cingulata, Haemaphysalis punctata, Helminthosporium* sp., *Ixodes redikorzevi, Lentius lepideus, Lenzites trabea, Meloidogyne incognita, M. javanica, Monilinia fructicola, Musca domestica, Myrothecium verriacaria, Pericallia ricini, Pestalotia* sp., *Phomopsis* sp., *Polyporus versicolor, Pseudomonas lachrymans, P. phaseolicola, P. solanacearum, Pseudoperonospora cubensis, Pyricularia oryzae, Rhipicephalus rossicus, Sphaceloma ampelinum, Spodoptera littoralis, S. litura, Trichoderma viride, Trogoderma granarium, Ustilago avenae, U. hordei, U. tritici, Verticillium alboatrum* and *Xanthomonas campestris.*

### *Annona reticulata*

*Achaea janata, Aphis fabae, Callosobruchus chinensis, C. maculatus, Coccus viridis, Crocidolomia binotalis, Dysdercus cingulatus, Epacromia tamulus, Euproctis fraterna, Hypsa ficus, Idiocerus* sp., *Leanium* sp., *Macrosiphoniella sanborni, Macrosiphum solanifolii, Pediculus humanus capitis, Plutella xylostella, Spodoptera litura* and *Tribolium casteneum.*

### *Chrysanthemum cinerariifolium*

*Acyrthosiphum pisum, Aedes aegypti, Anopheles quadrimaculatus, Antestiopsis lineaticollis, Aphis fabae, A. pomi, A. sorbi, A. spiraecola, Asphondylia* sp., *Autographa brassicae, Bombyx mori, Brevicoryne brassicae, Cladius pectinicornis, Diabrotica duodecimpunctata, D. punctata, Doryphora 10-lineata, Earias fabia, Empoasca devastans, E. fabae, Endelomyia rosae, Ephestia elutella, Epicauta pennsylvanica, Epilachna varivestes, Eriosoma tesselatum, Erythroneura comes, Eutettix tennellus, Galerucella luteola, Gargaphia solani, Gnorimoschema lycopersicella, Laspeyresia molesta, L. pomonella, Leptinotarsa decemlineata, Leucinodes orbonalis, Lipaphis erysimi, Lygus elisus, L. hesperus, Macrosiphoniella sanborni, Macrosiphum rosae, Macrosteles divisus, Malacosoma americana, Mamestra picta, Mineola scitulella, Murgantia histrionica, Musca domestica, Ophiomyia reticulipennis, Periphyllus lyropictus, Phadon cochleariae, Phyllotreta vittata, Pieris brassicae, P. rapae, Plodia interpunctella, Plutella xylostella, Pteronus ribesii, Pyrausta nubilalis, Sitophilus oryzae, Sitrotroga cerealella, Tetranychus cinnabarinus, Thermobia domestica* and *Toxoptera aurantii.*

### *Derris elliptica*

*Aphis fabae, A. medicaginis, Autographa brassicae, Bombyx mori, Busseola fusca, Callosobruchus chinensis, Ceratitis capitata, Coccus viridis, Crocidolomia binotalis, Epilachna varivestes, Euphydryas chalcedona, Hadena oleracea, Macrosiphum liriodendri, Malacosoma neustria, Meloidogyne incognita, Menopon biseriatum, Myzus persicae, Oryzaephilus surinamensis, Phalera bucephala, Phymatocera aterrima, Pieris brassicae, P. rapae, Plutella xylostella, Pteronus ribesii, Pyrausta nubilalis, Pyricularia oryzae, Rhopalosiphum persicae* and *Spodoptera litura*

### *Lantana camara*

*Aphis fabae, Athalia proxima, Dysdercus cingulatus, D. koenigii, Lipaphis erysimi, Manduca sexta, Musca domestica, Ostrinia furnacalis, Panonychus citri, Plutella xylostella* and *Sitophilus oryzae.*

### Mammea americana

*Andrector ruficornis, Ascia monuste, Attagenus piceus, Blatella germanica, Cerotoma ruficornis, Ctenocephalides canis, Culx sp., Diabrotica bivittata, Diaphania hyalinata, Laphygma frugiperda, Macrosiphum sonchi, Myzus persicae, Oncopletus fasciatus, Pachyzancla bipunctalis, Peridroma saucia, Periplaneta americana, Pieris rapae, Plutella xylostella, Prenolepis longicornis, Pseudaletia unipuncta, Rhipicephalus sanguineus, Sitophilus oryzae, Spodoptera eridania, S. frugiperda* and *Tineola bisselliella.*

### Ocimum sanctum

*Alternaria tenuis, Amaranthus spinosus, Curvularia penniseti, Dacus correctus, Drechslera oryzae, Dysdercus cingulatus, Exserohilum turcicum, Helminthosporium* sp., *Meloidogyne incognita, M. javanica, Musca domestica, Pericallia ricini, Pyricularia oryzae, Rhizoctonia solani* and *Spodoptera litura.*

### Piper nigrum

*Acanthoscelides obtectus, Anthonomus grandis, Callosobruchus maculatus, Dysdercus cingulatus, Fusarium oxysporum, Heliothis obsoleta, H. zea, Musca domestica, Plutella xylostella, Pseudomonas solanacearum, Sitophilus oryzae, S. zeamais* and *Tribolium castaneum.*

### Vitex negundo

*Achaea janata, Callosobruchus chinensis, Euproctis fraterna, Latheticus oryzae, Musca domestica, Pericallia ricini, Plutella xylostella, Pyricularia oryzae, Sitophilus oryzae, Sitotroga cerealella, Spodoptera litura* and *Tryporyza incertulas.*

### Zingiber officinale

*Drechslera oryzae, Rhizoctonia solani, Sclerotium oryzae, S. rolfsii* and *Tribolium casteneum.*

# Index